FUELS TO DRIVE OUR FUTURE

Committee on Production Technologies for Liquid
Transportation Fuels

Energy Engineering Board

Commission on Engineering and Technical Systems

National Research Council

NATIONAL ACADEMY PRESS
Washington, D.C. 1990

National Academy Press • 2101 Constitution Avenue, N.W. • Washington, DC 20418

NOTICE: The project that is the subject of this report was approved by the Governing Board of the National Research Council, whose members are drawn from the councils of the National Academy of Sciences, the National Academy of Engineering, and the Institute of Medicine. The members of the committee responsible for the report were chosen for their special competences and with regard for appropriate balance.

This report has been reviewed by a group other than the authors according to procedures approved by a Report Review Committee consisting of members of the National Academy of Sciences, the National Academy of Engineering, and the Institute of Medicine.

The National Academy of Sciences is a private, nonprofit, self-perpetuating society of distinguished scholars engaged in scientific and engineering research, dedicated to the furtherance of science and technology and to their use for the general welfare. Upon the authority of the charter granted to it by the Congress in 1863, the Academy has a mandate that requires it to advise the federal government on scientific and technical matters. Dr. Frank Press is president of the National Academy of Sciences.

The National Academy of Engineering was established in 1964, under the charter of the National Academy of Sciences, as a parallel organization of outstanding engineers. It is autonomous in its administration and in the selection of its members, sharing with the National Academy of Sciences the responsibility for advising the federal government. The National Academy of Engineering also sponsors engineering programs aimed at meeting national needs, encourages education and research, and recognizes the superior achievements of engineers. Dr. Robert M. White is president of the National Academy of Engineering.

The Institute of Medicine was established in 1970 by the National Academy of Sciences to secure the services of eminent members of appropriate professions in the examination of policy matters pertaining to the health of the public. The Institute acts under the responsibility given to the National Academy of Sciences by its congressional charter to be an adviser to the federal government and, upon its own initiative, to identify issues of medical care, research, and education. Dr. Samuel O. Thier is president of the Institute of Medicine.

The National Research Council was organized by the National Academy of Sciences in 1916 to associate the broad community of science and technology with the Academy's purposes of furthering knowledge and advising the federal government. Functioning in accordance with general policies determined by the Academy, the Council has become the principal operating agency of both the National Academy of Sciences and the National Academy of Engineering in providing services to the government, the public, and the scientific and engineering communities. The council is administered jointly by both Academies and the Institute of Medicine. Dr. Frank Press and Dr. Robert White are chairman and vice chairman, respectively, of the National Research Council.

This is a report of work supported by Contract DE-FG01-89FE61694/R between the U.S. Department of Energy and the National Academy of Sciences.

Library of Congress Catalog Card Number 90-60382
International Standard Book Number 0-309-04142-2

Cover photograph: COMSTOCK.

Copyright © 1990 by the National Academy of Sciences

Printed in the United States of America

COMMITTEE ON PRODUCTION TECHNOLOGIES FOR LIQUID TRANSPORTATION FUELS

JOHN P. LONGWELL (*Chairman*), Massachusetts Institute of Technology, Cambridge, Massachusetts
WILLIAM FISHER (*Vice Chairman*), University of Texas, Austin, Texas
SEYMOUR ALPERT, Electric Power Research Institute, Palo Alto, California
BRUCE BEYAERT, Chevron Corporation, San Francisco, California
ROBERT HALL, Amoco Corporation, Chicago, Illinois
ROBERT L. HIRSCH, ARCO Oil and Gas Company, Plano, Texas
PAUL R. KASTEN, Oak Ridge, Tennessee
FLYNT KENNEDY, Consolidation Coal Company, Library, Pennsylvania
ROY KNAPP, University of Oklahoma, Norman, Oklahoma
IRVING LEIBSON, Bechtel Group, Inc., San Francisco, California
ARTHUR E. LEWIS, Lawrence Livermore National Laboratory, Livermore, California
PHILLIP S. MYERS, University of Wisconsin, Madison, Wisconsin
RONALD A. SILLS, Mobil Research and Development Corporation, Paulsboro, New Jersey
DANIEL SPERLING, University of California, Davis, California
JAMES LEE SWEENEY, Stanford University, Stanford, California
JOHN M. WOOTTEN, Peabody Holding Company, Inc., St. Louis, Missouri

Liaison Members with Energy Engineering Board

DONALD B. ANTHONY, BP Exploration, Inc., Houston, Texas
GLENN A. SCHURMAN, Chevron Corporation, San Francisco, California
LEON STOCK, Argonne National Laboratory, Argonne, Illinois

Staff

JAMES J. ZUCCHETTO, Study Director, Committee on Production Technologies for Liquid Transportation Fuels
MAHADEVAN (DEV) MANI, Associate Director, Energy Engineering Board
MICHELLE E. SMITH, Administrative Assistant
PHILOMINA MAMMEN, Administrative Assistant

ENERGY ENGINEERING BOARD

JOHN A. TILLINGHAST (*Chairman*), Tiltec, Portsmouth, New Hampshire
DONALD B. ANTHONY, BP Exploration, Inc., Houston, Texas
RALPH C. CAVANAGH, Natural Resources Defense Council, San Francisco, California
CHARLES F. GAY, Arco Solar, Inc., Camarillo, California
WILLIAM R. GOULD, Southern California Edison Company, Rosemead, California
JOSEPH M. HENDRIE, Brookhaven National Laboratory, Upton, New York
WILLIAM W. HOGAN, Harvard University, Cambridge, Massachusetts
ARTHUR E. HUMPHREY, Lehigh University, Bethlehem, Pennsylvania
BAINE P. KERR, Pennzoil Company, Houston, Texas
HENRY R. LINDEN, Illinois Institute of Technology, Chicago, Illinois
THOMAS H. PIGFORD, University of California, Berkeley, California
MAXINE L. SAVITZ, Garrett Ceramic Component Division, Torrance, California
GLENN A. SCHURMAN, Chevron Corporation, San Francisco, California
WESTON M. STACEY, Georgia Institute of Technology, Atlanta, Georgia
LEON STOCK, Argonne National Laboratory, Argonne, Illinois
GEORGE S. TOLLEY, University of Chicago, Chicago, Illinois
DAVID C. WHITE, Massachusetts Institute of Technology, Cambridge, Massachusetts
RICHARD WILSON, Harvard University, Cambridge, Massachusetts
BERTRAM WOLFE, General Electric Nuclear Energy, San Jose, California

Technical Advisory Panel

HAROLD M. AGNEW, GA Technologies, Inc., Solana Beach, California
FLOYD L. CULLER, JR.,* Electric Power Research Institute, Palo Alto, California
KENT F. HANSEN,* Massachusetts Institute of Technology, Cambridge, Massachusetts
MILTON PIKARSKY,† The City College, New York, New York
CHAUNCEY STARR, Electric Power Research Institute, Palo Alto, California
HERBERT H. WOODSON, University of Texas, Austin, Texas

---

*Liaison members of the Commission on Engineering and Technical Systems to the Energy Engineering Board.
†Deceased June 1989.

Staff

ARCHIE L. WOOD, Director, Energy Engineering Board
MAHADEVAN (DEV) MANI, Associate Director, Energy Engineering Board
JAMES J. ZUCCHETTO, Senior Program Officer
KAMAL ARAJ, Senior Program Officer
ROBERT COHEN, Senior Program Officer
JUDITH A. AMRI, Administrative/Financial Assistant
MARY C. PECHACEK, Administrative Secretary
PHILOMINA MAMMEN, Administrative Secretary
THERESA M. FISHER, Administrative Secretary

Consultants

Norm Haller
George T. Lalos

Preface

The steady decline in the fraction of U.S. transportation fuels supplied from domestic resources coupled with the instability, price volatility, and increase in the gap between imports and exports resulted in a request by the U.S. Department of Energy (DOE) for preparation of this report by the National Research Council (NRC). The objective of this study is to outline for DOE a broad R&D program aimed at producing liquid transportation fuels from domestic resources. (See Appendix A for statement of task.)

In general, technologies are known for production from our major resources—oil, gas, coal, western oil shale, tar sands, and biomass; however, the cost is higher than the current cost of imported petroleum. In all of these technologies for production and conversion, there are, however, substantial opportunities for cost reductions.

While predictions of the future price of imported oil are unreliable, there is sufficient probability that prices will rise to the level where the new technologies can be economically applied within the next 20 years and that an effective R&D program in this area constitutes an important assurance for future supply at minimum price. The committee focused, therefore, on the R&D needed for cost reduction within this 20-year time frame. Since the environmental problems caused by the use of transportation fuels are also of increasing importance, R&D on these problems was also considered to be an important part of the overall transportation fuel supply R&D program.

Four meetings of the entire committee were held. One was combined with a 2-day workshop at which presentations were made by experts from government, industry, and academia and by committee members. These presentations made a major contribution to the background necessary for the report, and the efforts of all presenters are greatly appreciated (see

Appendix B). To assemble a full draft of the report, a fifth meeting was held by a subcommittee consisting of committee members William Fisher, Robert Hall, Roy Knapp, James Sweeney, and John Longwell and NRC staff members James Zucchetto and Dev Mani.

The rapid pace at which this task was completed called for a high level of participation and ability to quickly resolve differences in the members' viewpoints. The report thus represents the combined views of individual members of the committee but not necessarily those of the organizations employing them.

The committee's analysis was facilitated by the work of consultants. Velo Kuuskraa, Kathleen McFall, and Michael Godec of ICF Resources, Inc. (Fairfax, Virginia), summarized available information on U.S. reserves and resources of petroleum and natural gas. Bernard Schulman and Frank Biasca of SFA Pacific, Inc., performed a cost analysis of converting various feedstocks into transportation fuels. Their reports are available directly from the consultants.

Of special importance were the contributions of James Zucchetto, Senior Program Officer, whose efforts in organizing the committee and its activities were essential to the successful completion of this task.

JOHN P. LONGWELL, *Chairman*
Committee on Production Technologies
for Liquid Transportation Fuels

Contents

EXECUTIVE SUMMARY 1

1 INTRODUCTION... 10
    Objective of the Study, 10
    U.S. R&D for Liquid Fuels Production from Domestic
      Resources, 11
    Current Concerns About Energy and the U.S. Transportation
      System, 13
    Increasing the Use of Domestic Resources, 15
    Planning Scenarios, 17
    Organization of the Study and Report, 19

2 CONVENTIONAL PETROLEUM, ENHANCED OIL RECOVERY,
  AND NATURAL GAS 21
    Remaining Domestic Oil and Gas Resources, 24
    Production Technologies and Processes, 28
    Upstream Oil and Gas Environmental Impacts, 31
    Time and Investment Required for Increased Oil and Gas
      Production, 32
    Loss of Reserve Growth and EOR Potential, 33
    Technological Opportunities, 35
    DOE Research Program, 38
    Summary, 39

3 PRODUCTION COSTS FOR ALTERNATIVE LIQUID
  FUELS SOURCES 40
    Structure of the Analysis, 40
    Cost Estimates for the Various Technologies, 43
    Issues of Fuel Distribution and Use, 53
    Conclusions, 56

ix

4 CONVERSION TECHNOLOGIES AND R&D
   OPPORTUNITIES 57
   Production of Hydrogen and Synthesis Gas, 57
   Heavy Oil Conversion, 66
   Tar Sands Recovery and Processing, 69
   Oil Shale, 76
   Syngas-Based Fuels, 87
   Direct Coal Liquefaction, 92
   Coal-Oil Coprocessing, 97
   Coal Pyrolysis, 100
   Direct Conversion of Natural Gas, 102

5 ENVIRONMENTAL IMPACTS OF ALTERNATIVE FUELS 105
   Air Quality, Health and Safety Effects, 105
   Greenhouse Gas Emissions, 111

6 MAJOR CONCLUSIONS AND RECOMMENDATIONS FOR
   R&D ON LIQUID TRANSPORTATION FUELS 115
   Overview, 115
   Resources, 117
   Environmental Considerations, 122
   Major Conclusions and Recommendations, 123

APPENDIXES

A. Statement of Task... 133
B. Committee Meetings and Activities......................... 135
C. U.S. and World Resources of Hydrocarbons 138
D. Cost Analysis Methods 146
E. Technologies for Converting Heavy Oil 178
F. Retorting Techologies for Oil Shale 183
G. Research, Development, and Demonstration Definitions 185
H. Coprocessing Technology 188
I. Technical Data for Coal Pyrolysis 191
J. Description of Technologies for Direct Conversion of
   Natural Gas 197
K. Temperature Characteristics of High-Temperature Gas Reactors .. 200

GLOSSARY 201

REFERENCES AND BIBLIOGRAPHY 205

INDEX 213

FUELS TO DRIVE OUR FUTURE

Executive Summary

This report of the National Research Council's (NRC's) Committee on Production Technologies for Liquid Transportation Fuels addresses the problem of producing those fuels from domestic resources. Included in the report are an economic analysis of the various technologies, an assessment of their state of development, and suggested strategic directions for a 5-year R&D program for producing liquid transportation fuels from plentiful domestic resources (see Chapter 6 for a more detailed summary of the report).

In addition to conventional gasoline, diesel, and aviation fuels, there is growing interest in alternatives such as methanol and compressed natural gas. Concerted efforts are under way to understand better the consequences to human health, air pollution, and the greenhouse effect from the use of these fuels.

While this report concentrates on R&D important to fuels production from domestic resources with priority given to lowest cost, it is recognized that choice of fuels, the feedstock for their manufacture, and fuel composition will be strongly influenced by these additional considerations. Thus, the R&D program should be flexible enough to anticipate and accommodate changes that may be required for environmental and other reasons. This viewpoint is reflected in the committee's recommendations.

Analysis of these problems, however, was severely limited by the time constraint for completion of this study and by the study goals. It is conceivable that increasing concerns about global climate change could affect the balance of R&D expenditures on fossil vs. nonfossil energy technologies. Additional studies are needed and it is anticipated that the ongoing NRC study by the Committee on Alternative Energy R&D Strategies will make an important contribution to the development of federal energy R&D

programs with the goal of reducing greenhouse gas emissions in the production and use of fuels and electricity.

In the 1970s increasing imports of petroleum to the United States and rapidly rising oil prices stimulated U.S. public and private development of domestic resources of oil and other fossil fuels as replacements for imported petroleum. These efforts were sharply curtailed in the 1980s as a result of falling international oil prices. U.S. petroleum exploration and development are down substantially as are private R&D on oil recovery and the conversion of such resources as coal, oil shale, and tar sands into liquid transportation fuels or substitutes for petroleum. Domestic petroleum production has been in decline the past few years, and petroleum imports reached 50 percent of total consumption in July 1989; imports of crude oil and refined products are approaching 50 percent of consumption of hydrocarbon liquids. Also, more rapid deterioration of domestic oil production is certain under current conditions. Since fuels used for transportation in the United States are derived almost entirely from crude oil and natural gas liquids, any use of domestic resources for transportation fuels can help reduce petroleum imports.

Some anticipation of future conditions is required to plan an R&D program. The committee considered a number of scenarios to structure its thinking concerning the U.S. Department of Energy's (DOE) future R&D program for liquid transportation fuels. The economic scenarios considered were: (I) oil prices stay at $20/barrel for the next 20 years; (II) oil prices rise to about $30/barrel between 10 and 20 years from now; and (III) oil prices rise to about $40/barrel or greater between 10 and 20 years from now (all prices are in 1988 dollars). The committee believes that Scenario II is the most probable, while Scenarios I and III are less probable but likely to occur. In addition, the committee judges that the potential for continued price volatility is high under any scenario.

Two basic environmental scenarios were considered: (IV) aside from greenhouse gas emissions, increasingly stringent general emission, waste disposal, and fuel composition regulations are established during the next 20 years; and (V) because of worldwide concerns about climatic changes, policies to control U.S. greenhouse gas emissions are implemented. These two scenarios are not mutually exclusive. Consideration was also given to government policies that either encouraged domestic production or were neutral.

Not only is U.S. petroleum production declining, but the industry's emphasis is changing. The major oil companies are increasingly investing abroad, because costs are lower, the potential for successful large oil fields is higher, and some developing countries are offering special incentives to encourage development of their petroleum resources. In addition, small independent companies and individuals in the United States have declined in number and financial health.

The committee's economic and technical analysis of potential oil and gas production shows that increased prices and advanced technologies from expanded R&D can significantly increase U.S. reserves: Technology development can also reduce costs (see Chapter 2, Table 2-1). Such developments could allow the United States to partially offset current declining domestic petroleum production trends for many decades. During this time R&D on technologies for converting nonpetroleum resources into liquid transportation fuels has the potential for significant cost reductions. However, under expected market conditions, stimulating U.S. oil and gas production in the near-term will require government incentives for investment in as well as the support of R&D.

The committee also conducted a consistent economic analysis of technologies for converting domestic feedstocks other than petroleum (coal, oil shale, tar sands, natural gas, and biomass) into transportation fuels (gasoline, diesel, aviation, alcohols, compressed natural gas). The entire fuel cycle was considered, and alternative transportation fuels were compared to gasoline on a cost per barrel of oil equivalent (costs that would make fuel from the alternative resource just as expensive to the end user as gasoline from crude oil). Natural gas and oil prices were assumed to be coupled (see Appendix D for details) so that natural gas prices increased about 30 percent as much as oil prices; some calculations decoupling natural gas prices were also performed. All combinations of resources and conversion technologies considered are more costly than converting domestic petroleum, at current world prices, into gasoline and diesel fuel (see Chapter 3, Figure 3-2). It was assumed that these conversion plants would be built under a normal (not crash program) construction industry environment, and the costs used were for second- or third-generation (not pioneer) plants.

Domestic heavy oil conversion, solvent extraction of tar sands, direct liquefaction of coal, and compressed natural gas appeared to be the most economically attractive, with estimated costs below $40/barrel (1988 dollars, 10 percent real discount rate; all subsequent costs in this section are in the same terms). Costs were also calculated for a 15 percent discount rate, which increases costs by several dollars depending on the technology (see Chapter 3 and Appendix D).

Gasoline and diesel fuel produced from domestic natural gas, oil shale conversion, pyrolysis of most tar sands deposits, and methanol produced from domestic natural gas and underground coal gasification have a higher range of estimated costs. These different technologies are in different stages of development, and some estimates are firmer than others. The costs of methanol and liquids produced by indirect liquefaction are expected to be comparable. However, the relatively low prices of natural gas overseas ensure that the production of methanol or conventional fuels by indirect liquefaction from natural gas would occur outside the United States, barring government intervention.

Information available to the committee on the U.S. biomass resource base and the costs of conversion to liquid fuels suggests that biomass-derived fuels will cost more than those from fossil fuels. This information was, however, inadequate to allow as detailed an assessment as was done with coal and shale. Such an assessment should be undertaken by DOE with updated information. That aside, it is the committee's view that biomass could supply a substantial but limited fraction of the total requirements for liquid fuels, but that for some time to come, fossil fuels will continue to be dominant in the transportation sector.

With developments in technology, these costs can change. For example, over the past 10 years the estimated costs for direct liquefaction of coal have been reduced substantially. The committee made estimates of the potential for cost reduction if further development of these technologies occurs. The committee believes that vigorous R&D efforts on coal liquefaction and oil shale have potential to bring the costs down to the $30/barrel range: this might begin to make these technologies competitive with petroleum within 20 years under the price trends prescribed in Scenarios II and III. Developments in any of the conversion technologies could change the relative economics among the different options.

Environmental considerations are also extremely important (as outlined under Scenarios IV and V). Air and water quality can generally be controlled at a cost that is very dependent on the degree of cleanup required. If policies are implemented to restrict emissions of greenhouse gases, R&D will be needed to address this problem. Use of natural gas as a fuel or biomass (not using fossil fuel for its production and annually grown) as a feedstock would result in lower CO_2 emissions than coal combustion and liquefaction. Improvements could also come from developing nonfossil sources for the process heat used, for example, in the direct liquefaction of coal. Improved fuel economy, although not a topic of the current study, can have an important national impact by reducing imports and greenhouse gas emissions.

Also, hydrogen addition or carbon removal is needed to upgrade these fossil sources from low hydrogen-to-carbon (H/C) ratios to higher H/C transportation fuels. Hydrogen production from water is currently done by rejecting oxygen from water by reacting water with carbon-containing fuels. To eliminate the resulting carbon dioxide, water would have to be split by heat, electrolysis, or photolysis based on noncombustion sources of energy, such as solar or nuclear energy. These are more expensive than production of hydrogen or heat using fossil fuels with current technology.

In the United States there are a number of efforts under way to reduce ozone formation and particulate concentrations in urban areas by reducing vehicle emissions. Alternative fuels, such as methanol or compressed natural gas, may lead to ozone reductions relative to gasoline; however, the environmental effects of methanol, compressed natural gas, and hydrogen are uncertain. They also have different toxicity and safety issues associated with them. Reformulated gasolines may also be helpful in reducing ozone, as will improved engine design and vehicle emission controls. Environmentally driven constraints on fuel composition could have an important influence on the choice of conversion process and related R&D programs. In general, increasing environmental regulations, such as under Scenarios IV and V, will add to the costs of production and use of transportation fuels. For example, if the aromatic content of gasoline is reduced, costs for gasoline made from direct coal liquefaction could increase somewhat. Other composition changes would be needed to maintain octane number.

R&D strategy should explicitly recognize the high degree of uncertainty in the U.S. energy and environmental future. It is impossible to predict petroleum prices. Accordingly, the extent of U.S. oil (and gas) resource utilization will depend on prices, the nature of the resource, government actions, and technical developments. If domestic production cannot be held near current levels, U.S. dependence on petroleum imports will increase. Even if petroleum prices reach levels that make conversion of nonpetroleum resources competitive with crude oil, private investment may be slow to occur because of the risks associated with new technology, concerns of price volatility, and the residual effects of the twin oil price collapses of 1986 and 1988.

In the face of these energy, technical, and environmental uncertainties, a diverse and substantial federal R&D program could provide multiple options and insurance for future domestic production.

RECOMMENDATIONS FOR LIQUID FUELS R&D

The committee has used four criteria for deciding on the appropriateness of research areas in liquid transportation fuels for the DOE program. They are: (1) the possible timing of commercial application, Scenario II being considered the most probable course for oil prices; (2) potential size of the resource and application; (3) potential for cost reduction and acceptable environmental impact; and (4) the need for DOE participation, based on the extent of private sector involvement. The committee divided the research areas into those of major, medium, and modest funding: These categories apply to the relevance of the activities to the development of domestic production technologies for liquid transportation fuels. Some of these research areas might have different funding levels for other applications of fossil resources such as electric power generation or industrial process heat (see Chapter 6 for more details). The percentage of the total fossil fuel budget for liquid fuels is about 25 percent; most of the remainder is related to coal combustion with electric utility application. The ranking within category areas is not in priority order.

Under Scenario II the premise that oil prices reach $30/barrel within 10 to 20 years conforms with a target of $30/barrel for coal and oil shale through pilot projects and studies over the next 5 years. Under Scenario I the pace of the program could be slowed in comparison with Scenario II, whereas the more rapid price increase under Scenario III would call for a more rapid pace. Increased emphasis on curtailing greenhouse gas emissions would result in more emphasis on nonfossil sources of process heat and hydrogen, whereas increasing environmental constraints would lead to greater emphasis on environmentally related activities. The recommended areas of research as proposed are diverse and provide options in the face of the uncertainty that these scenarios encompass. If the economic and environmental situation changes, the program would need to be adjusted.

Major Funding Areas

The resource areas for high funding are domestic oil and gas, coal, and oil shale resources. These represent large domestic resources, with oil R&D also providing a means of achieving a significant U.S. production over a period of time when coal and oil shale technologies can be further developed. The cost reduction potential for converting these resources into liquid transportation fuels as well as the need for a DOE role also make them important areas. The committee has not made a detailed analysis of required federal funding for R&D activities for these resources. They are generally of major importance, and this should be reflected in the relative funding levels among these areas.

There is less need for DOE funding of R&D on conventional gas production since activity outside DOE is expected to continue and possibly increase, but DOE should continue its work on unconventional gas recovery. Significant funding and attention are also recommended for research related to fuel composition and its environmental and end-use consequences.

1. *Participation in R&D and Technology Transfer for Oil and Gas Production.* Significant additions can be made to U.S. oil reserves with development of advanced recovery technologies and greater understanding of complex reservoirs. The DOE program should be in balance with other energy R&D areas and pursued in coordination with industry, both independents and major oil companies, preferably with direct industry participation. An effective program of information and technology dissemination is needed.

2. *Production from Coal and Western Oil Shale.* These vast U.S. resources have the potential to be converted into liquid fuels in the $30/barrel of oil equivalent category. A goal should be established to reach this cost while satisfying environmental requirements. For coal liquefactions, pilot plant and engineering studies should be conducted during the next few years to confirm that this goal can be achieved. If so confirmed, the DOE should take the lead in working with industry to further develop the technology with the design of a larger pilot plant (500 to 1000 bbl/day). Initiation of construction would depend on a new assessment of oil availability and costs at that time.

The current shale program is too small compared with the coal liquefaction program and should be expanded. A field pilot plant should be built over the next 5 years to test advanced retorting technologies that can show the potential for meeting environmental requirements and for achieving the $30/barrel oil equivalent cost category.

Because manufacture of transportation fuels from both of these resources produces more carbon dioxide than processes based on oil, natural gas, or some biomass processes, a special effort should be made to identify possible opportunities, such as using nonfossil sources of energy (e.g., nuclear or solar based) for process heat or hydrogen production, for reduction in emissions of related greenhouse gases from the conversion of coal and oil shale into liquid transportation fuels.

3. *Environmental and End-Use Considerations.* There are many uncertainties remaining on the environmental and end-use impacts of using alternative fuels such as methanol and compressed natural gas. The DOE, other agencies, and the private sector should work together to develop a better understanding of these impacts and characterize different fuel-engine-emissions control combinations to provide guidance on emission goals, their impact on vehicle performance and cost, and how they will affect fuel formulation and fuel composition goals for R&D on production technologies.

Moderate Funding Areas

4. *Coal-Oil Coprocessing.* Coprocessing of heavy oils or residuum with coal may offer an opportunity for the introduction of coal as a refinery feedstock. It is expected to have rather limited application unless important synergisms between coal and oil occur. Funding of basic bench-scale research should be continued over the next 5 years to define the extent of synergism when coal and residuum are processed together, followed by a thorough economic analysis quantifying the impact of any synergism.

5. *Tar Sands.* The tar sands resource can potentially make an important domestic contribution to liquid fuels production, and a large fraction is government owned. Liquid transportation fuels can potentially be produced

from a portion of this resource at $25 to $30/barrel oil equivalent with a hydrocarbon extraction process. Further, there is little industry activity in this area. Over the next 5 years candidate processes should be evaluated, and, if promising, further development and demonstration in a field pilot plant should be undertaken.

6. *Petroleum Residuum, Heavy Oil, and Tar Conversion Processes.* The resource base is substantial, but these processes have been under intensive development in both the domestic and foreign petroleum industries. The DOE should support a laboratory program that provides basic information at the molecular level to augment the private sector effort.

7. *Biomass Utilization.* Use of some biomass resources is one pathway that can result in less net release of greenhouse gases than fossil fuels contribute. Biomass supply constraints and costs will probably require continued use of fossil fuel resources. Use of biomass to produce liquid fuels directly is of continuing interest; however, by integration of processing of biomass and fossil resources (e.g., by generating process hydrogen from biomass instead of coal), a greater reduction in CO_2 from the combined processes may be achievable (as suggested in recommendation 2). There is little industry activity in this area. It is, therefore, recommended that research and systems studies be conducted on the optimum integration of biomass with fossil fuel conversion processes as well as for stand-alone biomass conversion processes.

8. *Coal Pyrolysis.* The current DOE program is aimed at production of pyrolysis liquids and metallurgical coke and does not have a high priority for liquid transportation fuels. Coal pyrolysis combined with production of synthesis gas has the potential for increasing liquid yields in conversion processes for liquid transportation fuels. Since there is little privately funded R&D in this area, medium priority is placed on a program of basic pyrolysis research. Systems studies should investigate integrating pyrolysis with direct coal liquefaction.

Modest Funding Areas

9. *Processes for Producing Methanol, Methanol-derived Fuels, or Fischer-Tropsch Liquids from Synthesis Gas.* Synthesis gas (carbon monoxide plus hydrogen) can be made from such feedstocks as natural gas or coal and subsequently converted into hydrocarbon liquids or methanol. Industry is vigorously studying these processes, and production is expected to be outside the United States, where natural gas prices are low. These factors discourage DOE work in this area beyond fundamental and exploratory research.

10. *Direct Methane Conversion.* This process is being studied at the bench scale at various institutions, but potentially significant cost reduc-

tions have not been demonstrated. If breakthroughs are achieved, production would occur in foreign locations. DOE work should be limited to continuing fundamental research.

11. *Eastern Oil Shale.* Although widespread, most eastern oil shale is low grade, occurs in thin seams, has a high stripping ratio for mining, and is inherently more expensive than western shale. The committee judges that the economic use of this resource will occur much later than coal or western oil shale. Hence, no development is recommended at this time.

1

Introduction

OBJECTIVE OF THE STUDY

The objective of this study is to outline for the U.S. Department of Energy (DOE) a broad 5-year R&D program aimed at producing liquid transportation fuels from plentiful U.S. energy resources (see Statement of Task, Appendix A). In a 1988 conference report, Congress enunciated concerns about increasing U.S. dependence on imported petroleum and the expected predominance of petroleum-based liquid fuels in transportation for the foreseeable future; it was also noted that improvements in coal liquefaction technologies had reduced costs (U.S. Congress, 1988a).

R&D for producing liquid transportation fuels should have several goals: greater use of domestic resources relative to crude oil imports; technologies that are viable in the face of changing conditions, such as environmental constraints; protecting the United States against the vagaries of the world oil market; and strengthening U.S. energy R&D and international competitiveness.

The choice of liquid transportation fuels in the United States is relatively narrow: gasoline, diesel, methanol, ethanol, aviation fuels, and liquid petroleum gas; the committee also briefly addressed compressed natural gas. The environmental implications of transportation fuels are becoming increasingly important in the debate on urban air quality. The committee considered a variety of plentiful domestic resources that could serve as feedstocks for these fuels: crude petroleum amenable to advanced recovery, heavy oils, tar sands, oil shale, coal, natural gas, and biomass. In identifying R&D directions the committee addressed the relative costs of various conversion technologies and identified opportunities for cost reduction through R&D.

INTRODUCTION

The goals of reducing petroleum imports and using domestic resources for transportation fuels require consideration of broad public policies yet to be resolved, a subject outside the scope of the committee's charter. For example, the federal government could create financial incentives for greater investment in domestic exploration and production. Such issues are not addressed by this study. Recommendations are made without the benefit of a consistent and broad U.S. energy policy, of which the production of liquid transportation fuels is only one aspect. The committee addressed R&D for production technologies with the notion that the increased use of U.S. resources could make a major contribution toward limiting the growth of petroleum imports.

U.S. R&D FOR LIQUID FUELS PRODUCTION FROM DOMESTIC RESOURCES

The DOE's Office of Fossil Energy is the primary federal organization that sponsors R&D directed toward producing and using fossil fuels (Table 1-1). According to congressional testimony on the Office's 1990 fiscal year budget, about $110 million of the 1989 budget, or 29 percent, was related in some way to liquid fuels. The petroleum budget included R&D on petroleum production as well as oil shale conversion. R&D on coal, however, involved coal combustion to a significant degree, although some of this research applied to producing liquid fuels. For example, developments in coal preparation can benefit coal liquefaction plants, and developments in coal gasification can be used in manufacturing hydrogen needed for direct or indirect liquefaction of coal.

In addition to the above programs in the DOE's Office of Fossil Energy, the DOE's Office of Conservation and Renewable Energy sponsors a $13 million/year program on biofuels energy technology. A major part of this work is applicable to liquid transportation fuels production.

The major oil companies continue to invest in the development of petroleum resources, but the twin oil price collapses of 1986 and 1988 led to significant drops in domestic exploration and development (Megill, 1989; U.S. Congress, 1987). In the past 5 to 7 years there has been a dramatic decline in industrial R&D on using domestic coal and oil shale for the production of liquid transportation fuels. Only a few companies now appear to have active programs in this area, and some of them are subsidized by federal funds. These efforts include the following:

- Union Oil is operating a commercial demonstration of oil shale retorting technology with a federal subsidy.
- Exxon has announced that it is continuing to develop coal liquefaction and oil shale retorting technologies with its own funds.

TABLE 1-1 DOE's Office of Fossil Energy R&D Program Budget (current dollars in millions)

	FY 1988 Appropriations	FY 1989 Appropriations	FY 1990 Request	FY 1990 House	FY 1990 Senate Panel
Coal Budget					
Control technology and coal preparation	$43.62	$48.93	$32.26	$60.10	$53.13
Advanced technology R&D	24.94	25.56	25.54	26.18	29.32
Coal liquefaction	27.13	32.39	9.66	37.68	33.26
Combustion systems	25.17	26.70	15.77	35.27	30.17
Fuel cells	34.20	27.53	6.50	38.40	29.80
Heat engines	17.95	22.83	8.92	20.02	21.22
Underground gasification	2.78	1.37	0.43	0.43	0.83
Magnetohydrodynamics	35.00	37.00	0	42.90	37.00
Surface gasification	22.99	21.56	8.74	19.64	29.88
Total coal	$233.78	$243.87	$107.82	$280.62	$264.61
Petroleum Budget					
Enhanced recovery	$16.54	$23.58	$18.24	$27.59	$28.46
Advanced process technology	3.43	4.20	4.62	3.60	3.60
Oil shale	9.50	10.53	1.68	8.18	10.88
Total oil	$29.47	$38.31	$24.54	$39.37	$42.94
Gas Budget					
Unconventional gas	$10.53	$11.38	$4.07	$13.17	$15.82
Cooperative R&D Ventures	$0	$0	$0	$4.80	$4.80
Total gas	$10.53	$11.38	$4.07	$17.97	$20.62
Miscellaneous[a]	$53.22	$88.03	$26.15	$84.72	$81.17
Total fossil R&D	$327.00	$381.59	$162.58	$422.68	$409.34

[a]Includes plant and capital equipment, program direction, environmental restoration, fuels conversion, and past year's offsets. Numbers may not add due to rounding.

SOURCE: July 31, 1989, Clean-Coal/Synfuels Letter.

- Amoco and Kerr-McGee are participating in DOE's direct coal liquefaction program using the test facility at Wilsonville, Alabama.
- Occidental Petroleum is proposing a cooperative program with DOE to demonstrate a modified in situ oil shale retorting process.
- The New Paraho Shale Oil Company is continuing the development of Paraho retorting technology. Their initial product focus is an asphalt-paving additive that may justify near-term commercialization. This product is being tested with government funds.
- Texaco and Dow Chemical have commercialized coal gasification technologies, both with federal assistance. Shell Oil is operating a demonstration unit using coal gasification technology. Although not the initial intended use, these technologies can be used as the first step of an indirect coal liquefaction facility and hydrogen generation for the production of liquid transportation fuels.

If these coal and oil shale technologies need to be ready for commercialization in the next 10 to 20 years, the federal government will be required to play a major role in furthering the RD&D efforts.

CURRENT CONCERNS ABOUT ENERGY AND THE U.S. TRANSPORTATION SYSTEM

In 1988 the U.S. transportation sector accounted for 63 percent of total U.S. petroleum consumption and will depend almost completely for the foreseeable future on liquid fuels for spark ignition (7.26 MMbbl/day of gasoline) and diesel engines (1.26 MMbbl/day of diesel fuel) and for air transport (1.04 MMbbl/day of jet fuel) but not for electric trains and some pipelines (EIA, 1989a). Although some efforts are under way to use compressed natural gas and electric vehicles in some urban areas, any transition from the use of liquid transportation fuels is likely to be slow.

Dependence on Imported Petroleum

There are concerns about the increasing U.S. dependence on imported petroleum, because the economy becomes more vulnerable to political events in oil-producing regions of the world and because the economic costs of these imports continue to increase. The desire for increased supply security has motivated military and political involvement in those regions of the world deemed vital to national security. However, sources of oil have diversified, and Saudi Arabian oil accounts for the maximum from a single country with 1.062 MMbbl/day in 1988: Total Arab OPEC oil accounted for 1.828 MMbbl/day. Establishment of the U.S. Strategic Petroleum Reserve of about 550 MMbbl has also reduced this vulnerability.

World oil prices rose substantially in the 1970s and early 1980s—stimulating conservation and oil exploration and production worldwide—but world oil prices plummeted in 1986 and 1988. These drops in prices have led to a significant decline in exploration, development, and production in the United States. U.S. crude oil production has fallen rapidly in the last few years, with 8.68, 8.35, and 8.18 MMbbl/day in 1986, 1987, and 1988, respectively (EIA, 1989a). Total production of petroleum liquids has also fallen from 10.9 MMbbl/day in 1986 to 10.51 MMbbl/day in 1988, while crude oil imports have risen from 4.18 MMbbl/day to 5.12 MMbbl/day. At the same time total consumption of oil products has risen steadily, reaching about 17 MMbbl/day in 1988. In July 1989 imports of petroleum products were 8.6 MMbbl/day, while total demand was about 17 MMbbl/day; thus, imports represented more than 50 percent of total demand (API, 1989). Annual net imports were also projected to increase in 1989.

The Energy Information Administration (EIA) anticipates that these trends will continue at least through the year 2000 (EIA, 1989a). EIA's base case forecast, assuming rising petroleum prices in the year 2000 to $28/barrel (in 1988 dollars) and enhanced vehicle efficiency, indicates that net imports (10.2 MMbbl/day) will become 55 percent of total U.S. consumption (18.6 MMbbl/day) by the year 2000. This situation could be more extreme if world oil prices remain low, further reducing domestic production and increasing demand.

U.S. proved petroleum reserves of about 27 billion bbl pales in comparison to the rest of the world's total of about 860 billion bbl (EIA, 1989a). However, U.S. total petroleum resources are quite extensive and could, if vigorously exploited, replace these reserves for a number of decades (see Chapter 2). Using these resources would require some combination of higher prices, advanced technologies, and government incentives.

Local Air Quality

Local air quality has become an important issue because many U.S. metropolitan areas fail to comply with national ambient ozone standards. Gasoline vehicles contribute to ozone formation by emitting volatile organic carbon compounds and oxides of nitrogen. The issue has entailed calls for the use of fuels different from gasoline. In some areas, such as the Los Angeles Basin of California, alternative fueled vehicles (e.g., those using methanol, natural gas, or electricity) are being investigated to determine their potential air quality effects. In other areas oxygenated gasolines are being required to reduce wintertime carbon monoxide emissions. U.S. diesel particulate standards have been promulgated for 1991 and 1994 and may necessitate the use of alternative fuels in some diesel engines (e.g., urban buses). Efforts are also under way to reformulate gasoline to facilitate improved vehicle emission control systems. These trends may affect the types and quantities of future transportation fuels.

Global Warming

Another major environmental issue is increasing man-made emissions into the atmosphere of carbon dioxide, nitrous oxide, methane, chlorofluorocarbons, and other gases, now collectively referred to as greenhouse gases because of their potential global warming effect. There are many uncertainties about the extent, rate, and potential impact of global warming. If greenhouse gases do in fact cause significant adverse changes in global weather patterns, global cooperation may be necessary to reduce these emissions and change future transportation fuels and the technologies and feedstocks for their production.

INCREASING THE USE OF DOMESTIC RESOURCES

By virtue of its mandate, the committee addressed opportunities on the supply side for converting domestic resources into liquid transportation fuels. Recent sustained and volatile low crude oil prices in world markets have resulted in decreased U.S. production. Private sector investment in the conversion of U.S. nonpetroleum resources into liquid transportation fuels is therefore constrained by the high costs relative to oil prices. Even if oil prices rose, the private sector would still hesitate to invest in such technologies given the uncertainties of future oil prices. The committee recognizes that fuel efficiency improvements would reduce import dependence, but such matters were not part of this study. Environmental concerns may also affect the balance of domestic production and imports because these partly determine the desirability of various fuel conversion technologies and their feedstocks. Proposals have also been discussed to convert remote gas in various regions of the world into methanol for transportation (U.S. DOE, 1988). This might diversify U.S. supplies but not solve the import problem.

U.S. petroleum and natural gas reserves are functions of prices, extraction technology costs, and potential environmental impact (Chapter 2, this volume; Kuuskraa et al., 1989; AAPG, 1989a,b). Proved reserves of petroleum at current conditions are about 27 billion bbl. Future potentially recoverable oil estimates vary widely depending on crude oil prices and technology. Table 1-2 presents petroleum reserve estimates for $24/barrel and $40/barrel with advanced technology (these figures include heavy oil and tar sands exploitation; see also Table 2-1 for a broader range of estimates). Natural gas reserve estimates for $3.00 per thousand cubic feet (Mcf) (1988 dollars at the wellhead) range from about 600 trillion cubic feet (Tcf) to 800 Tcf; at $5.00/Mcf reserves increase to between 900 and 1400 Tcf. (Note that in 1988 U.S. consumption of petroleum products was about 6.2 billion bbl, U.S. natural gas consumption was about 18 Tcf [3.25 billion bbl oil equivalent], and U.S. coal consumption was about 870 million short tons [3.37 billion bbl oil equivalent] [EIA, 1989a]).

TABLE 1-2 Estimates of U.S. Hydrocarbon Resources and Reserves

Resource	Approximate Amounts
Petroleum[a]	
Total resource base	484 billion bbl
Proved reserves	27 billion bbl
Recoverable at $24/barrel	76 to 106 billion bbl
Recoverable at $40/barrel	97 to 143 billion bbl
Natural Gas	
Proved reserves[b]	About 360 Tcf or 65 billion bbl oil equivalent (boe)
For $3.00/Mcf	From 600 to 900 Tcf or 110 to 160 billion boe
For $5.00/Mcf	From 800 to 1400 Tcf or 140 to 250 billion boe
Oil Shale[c]	
Western	560 to 720 billion bbl
Eastern	65 billion bbl
Coal[d]	
Geological resource	2.57 trillion tons or about 9.8 trillion boe
Demonstrated reserve base	490 billion tons or about 1.9 trillion boe

[a]Kuuskraa et al. (1989). Includes conventional petroleum, heavy oil, and tar sands. Proved reserves also include heavy oil. Recoverable numbers are for implemented and advanced technologies at the prices indicated. Also see Table 2-1.

[b]The lower end of the resource estimates is based on existing extraction technology and the higher end on advanced technology. See Table 2-2.

[c]See Appendix C and section on oil shale in Chapter 4.

[d]EIA (1984).

The U.S. resource base (including "in-place" and resources ultimately recoverable at some unspecified price) of heavy oil (oil with a viscosity between 100 and 10,000 cp) not yet produced is about 85 billion bbl, most of it in California and Alaska. The largest tar sands deposits (a bitumen deposit with greater than 10,000 cp in situ viscosity) occur in Utah and Alaska, with smaller deposits in Alabama, Texas, California, and Kentucky. Measured and speculative resources are estimated at about 22 billion and 41 billion bbl, respectively, for a total of about 63 billion bbl. (About 54 billion bbl represent deposits of 100 MMbbl or more.)

The largest and richest oil shale (an impure marlstone consisting of silicate and carbonate rock and the organic constituent kerogen) deposit is in the Piceance Creek Basin (Colorado), a part of the Green River formation that occurs in Colorado, Utah, and Wyoming. The bulk of the resource is on government land. Thinner and lower-grade shales occur in the eastern United States. Theoretical conversion of in-place western oil shale resources ranges from 560 billion to 720 billion bbl; eastern shale resources are estimated at around 65 billion bbl (Lewis, 1980; Riva, 1987).

The U.S. geological resource of coal is immense and widely distributed, and the reserve base that is technically recoverable is about 490 billion tons (1.9 trillion bbl of oil in terms of energy equivalent). Full development of underground gasification could extend this resource base.

There is considerable uncertainty regarding the biomass resource base in the United States. One estimate is that biomass resources might yield, as a maximum, assuming minimal disruption of the agricultural and silvicultural industries, 1 billion bbl/year (3 MMbbl/day) of hydrocarbon fuel equivalents made up of methanol, ethanol, vegetable oils, and others (Sperling, 1988). Emerging estimates suggest greater potential in the resource base than indicated in the above reference. The estimates and the economics of their conversion ought to be fully assessed by DOE, particularly with regard to increases in costs and impacts on agriculture and the environment as the resource base is fully utilized. Information that was available to the committee on the practical limits to the production of biomass fuels (i.e., without undue stress on the environment and on agriculture) indicate that biomass could supply a substantial but limited fraction of the total requirements for liquid fuels. As such, fossil fuels will continue to be dominant in the transportation sector for some time to come.

Conversion of most nontraditional domestic resources into liquid transportation fuels is not economic at current world crude oil prices of about $20/barrel (average refiner acquisition costs in 1988 dollars) or less but could become competitive as oil prices rise.

The DOE's R&D program should pursue those approaches that may enable the economic exploitation of domestic energy resources to produce liquid transportation fuels. The emphasis and pace of such a program ought to be conditioned by the national need to adjust to increased environmental constraints and uncertain world oil prices and by the nature and extent of R&D that might be undertaken independently in the private sector.

PLANNING SCENARIOS

To address the formulation of DOE's program, the committee considered different planning scenarios, with attention to world oil price trends, environmental concerns, and national security policy. These scenarios are not intended as forecasts but provide a context for planning a federal R&D program. In evaluating the R&D needs for transportation fuels, the committee compared the attractiveness of different technological options in light of these various possible scenarios.

Economic Scenarios

World oil prices have been the primary determinant of U.S. liquid fuel prices for decades. This connection is likely to hold for the foreseeable future. Recent experience exemplifies both the uncertainties about price trends and the unreliability of predictions. Although energy economists have learned important lessons from the past, they cannot predict oil prices with confidence. One school of thought suggests that future world oil demand will grow, leading to increasing OPEC control of the market with prices rising above current levels. Another school of thought believes that substitutes for OPEC oil production, including oil and natural gas from non-OPEC nations, could induce relatively low and stable oil prices. The committee has not ruled out either school of thought. Since meaningful planning depends on projections of possible future world oil prices, it must address a wide range of projections.

The EIA annually prepares three plausible projections of energy prices extending through the year 2000 (EIA, 1989a). The most recent forecasts present three world oil price trajectories reaching, for the year 2000, about $22, $28, and $35/barrel of crude oil (in 1988 dollars) imported to U.S. refiners (see Table D-2 in Appendix D). For investment and R&D planning a time horizon greater than 10 years is needed. For purposes of this study the EIA study results were extended as follows:

Scenario I. Future world oil prices remain less than or equal to about $20/barrel (in 1988 dollars) over the next 10 to 20 years.

Scenario II. Future world oil prices rise to about $30/barrel (in 1988 dollars) over the next 20 years.

Scenario III. Future world oil prices rise to about $40/barrel or greater (in 1988 dollars) over the next 20 years.

Crude oil prices are likely to be volatile, owing largely to war, revolution, or other political instability; to actions by exporting countries; and to the delicate supply-demand market balance easily upset by temporary ups and downs in oil output and in buyer expectations. Significant price excursions could be expected to occur within each of these scenarios notwithstanding their long-term trends. Yet while actual prices may significantly influence industry investment decisions for producing liquid fuels, such variations should not perturb the federal R&D program on a short-term basis. The government program should be based on long-term fundamentals.

Environmental Scenarios

Environmental concerns relating to liquid fuel production, transportation, and end use are complex and increasingly significant to both U.S. citizens and the federal government. Many of these concerns can be dealt with by employing various improved control technologies and increased safeguards against accidents; however, control technologies increase both capital and operating expenses.

Concerns about greenhouse gases resulting from production and use of hydrocarbon fuels force consideration of feedstock choices and alternative sources of heat and energy for fuel manufacture. For R&D planning the following two scenarios were chosen to reflect actions that might be taken to protect the environment. Scenario IV does not include greenhouse gases, and the two scenarios are not necessarily mutually exclusive.

Scenario IV. Much more stringent general emissions and waste disposal regulations are established.

Scenario V. Controls are placed on U.S. emissions of greenhouse gases because of concerns about global climate change.

Energy Security Scenarios

The last planning dimension is government policy on energy security. In this area various issues must be considered, particularly national security concerns about the growing dependence on foreign oil. For simplicity the committee chose two scenarios for the purpose of this study:

Scenario VI. Government policies neither encourage nor discourage liquid fuel production from domestic resources.

Scenario VII. Government policies encourage liquid fuel production from domestic resources.

The implications of these scenarios for a federal R&D program on conversion technologies for liquid transportation fuels are discussed in Chapter 6.

ORGANIZATION OF THE STUDY AND REPORT

In evaluating the various conversion technologies and appropriate R&D that should be pursued, the committee was helped in several ways. Consultants from ICF Resources, Inc. (Fairfax, Virginia), summarized estimates of economically recoverable U.S. reserves and resources of petroleum and natural gas with committee guidance (Kuuskraa et al., 1989). Chapter 2 of this report addresses U.S. oil and gas resources, the R&D that can reduce costs and improve production, and the relationship of prices to U.S. reserves and supply. Costs reported in the literature and potential cost reductions for converting various resources into transportation fuels were summarized and adjusted to a comparable basis by SFA Pacific, Inc. (Mountain View, California) (Schulman and Biasca, 1989). These cost estimates served as a basis for the committee to continue the analysis and refine the costs. A workshop was also held to gather information from a wide variety of experts. Chapter 3 presents the cost results, including some of the end-use issues, such as those related to alternative-fueled vehicles and fuel distribution, that can affect the costs of various fuels.

Chapter 4 addresses the present state of conversion technologies, environmental issues related to production, and R&D directions for the DOE. Chapter 5 addresses the environmental implications of using alternative fuels. Chapter 6 provides an integrating discussion and suggests general directions for a 5-year DOE R&D program in the context of the scenarios considered.

2

Conventional Petroleum, Enhanced Oil Recovery, and Natural Gas

Over geologic time oil and gas have accumulated in porous rock formations called reservoirs, where they are hydrodynamically trapped by overlying and adjacent impermeable rock. The origins of these hydrocarbons are generally believed to be plant and animal life buried millions of years ago and slowly transformed by pressure and temperature into oils and gases of various qualities. The oil or gas resides together with varying amounts of water in microscopic pore spaces within the reservoir rock.

Reserves are the amounts of oil or gas believed to be economically recoverable from a reservoir through the use of existing technology. This seemingly simple concept is, in fact, quite complicated. When a discovery well penetrates a new reservoir, it provides very little information about the complex geologic character of the reservoir and its contained fluids. Only after extensive drilling and production over time can the extent of the reserves be estimated with significant accuracy.

Production rates from reservoirs depend on a number of factors, such as reservoir pressure, rock type and permeability, fluid saturations and properties, extent of fracturing, number of wells, and their locations. Operators can increase production over that which would naturally occur by such methods as fracturing the reservoir to open new channels for flow, injecting gas and water to increase the reservoir pressure, or lowering oil viscosity with heat or chemicals. These supplementary techniques are expensive, and the extent to which they are used depends on such external factors as the operator's economic condition, sales prospects, and perceptions of future prices.

The extraordinary geological variability of different reservoirs means that production profiles differ from field to field (for illustrative purposes see Figure 2-1). Oil reservoirs can be developed to significant levels of

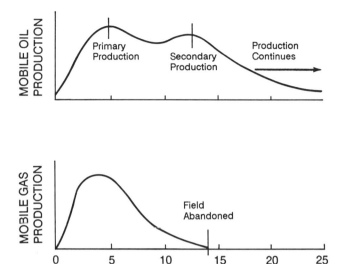

FIGURE 2-1 Production profiles of typical oil and gas reservoirs whose production is not limited by market constraint or government regulation. Oil flows generated by natural geological pressures are called primary production. Secondary production results from supplementary actions, such as water injection into the reservoir.

production and maintained for a period of time by supplementing natural drive force, while gas reservoirs normally decline more rapidly. On this basis an oil reservoir with the seemingly large reserve of a million barrels might produce only 200 to 400 bbl/day during its best years. Against a U.S. consumption of roughly 17 MMbbl/day of oil, that is indeed only a modest contribution. Nevertheless, a very large number of small U.S. reservoirs account for a significant part of domestic petroleum production.

The bulk of low-cost worldwide oil reserves is located in countries belonging to the Organization of Petroleum Exporting Countries (OPEC) (EIA, 1989a). Furthermore, because the marginal cost of production in OPEC is lower in comparison to costs in other countries, OPEC could exert major influences on oil prices. However, the United States has significant petroleum resources and an oil industry infrastructure that has shown it is capable of replacing reserves when the economic climate stimulates it to do so.

The U.S. oil and gas industry can be divided into upstream operations, which are concerned with exploring for and producing oil and gas, and downstream operations, which deal with transportation, refining, distribution, and marketing. Upstream operations were hardest hit by the 1986 and 1988 world oil price collapses.

In 1985 the U.S. upstream sector was a huge economic enterprise employing more than 600,000 people with sales of well over $500 billion (*API Basic Petroleum Data Book 1986; Oil and Gas Journal*, 1986). Participants included the domestic operations of a dozen major oil companies with 1985 net incomes of more than $15 billion; over 20,000 independent oil and gas explorers and producers; hundreds of oil field service companies who provided seismic surveys, drilling, logging, fracturing, and other services; and a large array of suppliers of pipe, pumps, compressors, computers, chemicals, and other equipment and supplies (*Oil and Gas Journal*, 1986). Currently, this enterprise is dramatically smaller in size.

The search for oil and gas is risky and expensive. For example, a prospect with a million barrels of reserve, which is considered to be relatively small "stakes," might cost several million dollars for people, land, geological and geophysical surveys, drilling, and testing. The likelihood of success—the "chance factor"—of such a relatively modest endeavor might range typically from 10 to 40 percent depending on the geographic proximity to other production. Chance factors for higher-stakes, larger targets can range from 5 to 20 percent. The most recent, famous, expensive high-stakes exploration failure was the 1984 Mukluk prospect in the Alaskan Beaufort Sea. It cost about $1.5 billion, and no commercial oil or gas was discovered.

Although predrilling geological and geophysical techniques for finding oil and gas have become increasingly more sophisticated in recent years, they are still far from providing clear images of the subsurface, including identifying the presence of oil and gas. There is no substitute for drilling a well to confirm geological models and to establish the presence of hydrocarbons. On this basis the number of drilling rigs operating at any particular time—the "rig count"—is an important measure of U.S. upstream activity.

There are different kinds of wells drilled in upstream operations. A development well is drilled to enable increased production from a known reservoir. It is not planned to increase reserves; however, it generally increases hydraulic communication with the reservoir units and often results in reserve appreciation. In geologically complex reservoir systems, targeted infill (drilling within a known field) development wells can significantly increase reserves. An extension exploration well is aimed at finding a new reservoir near an area of known production. An extension well can have a relatively high chance of success—30 to 60 percent—because of its proximity to known production.

Finally, there is wildcat exploration drilling. This is done in areas where there may be no subsurface information from existing wells. Wildcat exploration is the most risky drilling. During the two decades from 1960 to 1979, only 1 to 2 percent of the U.S. wildcat wells yielded significant new fields with reserves greater than 1 MMbbl of oil or 6 Bcf of gas; however,

10 to 16 percent of such wells yielded enough oil or gas to be brought into commercial operation. Thus, while rig count is a significant measure of industry activity, it is important to understand the types of holes being drilled. If the wells being drilled are developmental ones, the national reserve base will soon be more rapidly decreased as production from these wells depletes reserves. If there is a significant amount of exploration and targeted extension drilling, the national reserve base will be supplemented with additions. The recent decline in U.S. drilling activity, coupled with a shift to development drilling, will assuredly lead to a dramatic reduction in U.S. oil and gas production in the years ahead. Unless this is intentionally changed by increased domestic drilling and improved recovery technology, the United States can expect an increasing reliance on imported petroleum.

REMAINING DOMESTIC OIL AND GAS RESOURCES

Over time, judgments of volumes of U.S. oil and gas available for discovery, further development, and recovery have changed. Estimates of remaining resources made from 1900 to 1940 later proved to be exceedingly low. The sustained discovery of giant oil and gas fields from the 1920s through the middle 1950s led to more optimistic estimates of ultimate discovery and recovery. However, by the 1950s most of the onshore giant fields in the lower 48 states had been discovered and new field discovery of oil and natural gas had peaked. This peaking, preceded by an exponential increase in discovery, led Hubbert (1962) to postulate a symmetrical life cycle of both discovery and subsequent production. He believed that after peaking, discovery rates and, later, production rates would decline exponentially. Physically constrained, technology and economics could modify only slightly the volume of ultimate recovery. The volume of ultimate recovery was estimated using this model, and it was projected that production in the continental United States would peak in the early 1970s. When production did peak in the early 1970s and then declined through the rest of the decade, the remaining U.S. oil and gas resource was generally perceived to be shrinking and depleting rapidly.

Significant increases in wellhead prices for oil and natural gas through the 1970s and early 1980s led to a major increase in drilling. During this time approximately 45 percent of all gas wells and 30 percent of all U.S. oil wells were drilled and completed. This major pursuit of a resource base, which was judged marginal and rapidly depleting, showed unexpected response. Finding rates, expressed in volumes of oil and gas discovered per foot of exploratory drilling, though lower than those of the early days of exploration, remained stable. The rates did not decline exponentially with cumulative drilling as anticipated. Reserve growth from the extensive and intensive drilling of many older oil fields, generally judged to be fully appreciated by historical field development, was much greater than expected. By 1985 production in the continental United States was nearly 7.2 MMbbl/day—2 MMbbl/day more than projected using the exponential decline model. While this had the appearance of only arresting decline and stabilizing production, it actually represented a nearly 40 percent increase over projected production capacity from a resource once judged to be nearly exhausted. Estimates of ultimate recovery and production in the lower continental United States, earlier reported using the symmetrical life-cycle view of resource behavior, have already been exceeded.

Contrary to conventional wisdom, the growing view in 1989 was that the remaining U.S. resource base of oil and gas is substantial and sufficient to slow the rate of decline in production and maintain significant oil production for three to five decades at moderate and stable real prices. Remaining U.S. resources, like the resources pursued in recent years, will generally be converted to producible reserves in small to moderate-size increments. Remaining large reserve increments are largely restricted to the relatively high-cost frontier areas, including Alaska and the deep water offshore, attractive areas for the industry to explore.

Oil Resources

Two recent estimates of remaining U.S. oil resources have been made: one by the American Association of Petroleum Geologists (AAPG, 1989a) and the other by ICF Resources in their report to this committee (Table 2-1) (Kuuskraa et al., 1989). The ICF analysis incorporated estimates by the U.S. Department of the Interior (DOI) (U.S. Geological Survey [USGS] for onshore and the Minerals Management Service [MMS] for offshore) of undiscovered resources in 1989 (U.S. DOI, 1989).

Differences in proved reserves and reserve growth resource estimates made by AAPG and ICF Resources under assumptions of moderate cost ($25/barrel, 1986 dollars, assumed by AAPG; $24/barrel, 1988 dollars, assumed by ICF), moderate cost with advanced technology, and high cost ($50/barrel, 1986 dollars, assumed by AAPG; $40/barrel, 1988 dollars, assumed by ICF) are very slight when adjustments are made for the different real price assumptions. With the high-cost and advanced technology category, the difference in the two estimates is substantial, even allowing for the 38 percent difference in price assumptions. Obviously there is no historical experience with prices at either of the high-cost levels assumed. When this is combined with assumptions about advanced technology, the range of uncertainty is expectedly high.

The principal difference in the two estimates lies in judgments of volumes of undiscovered oil at the assumed price levels. The larger volumes

TABLE 2-1 Recent Estimates of U.S. Oil Resources (billion barrels)

	AAPG[a] (1989)	ICF Resources[a] (1989)
Moderate Costs		
Proved reserves	27.0	27.3
Reserve growth[b]	17.0	35.9
Undiscovered	33.0	12.4
Total	77.0	75.6
Moderate Costs with Advanced Technology		
Proved reserves	27.0	27.3
Reserve growth	62.0	69.2
Undiscovered	40.0	18.5
Total	129.0	115.0
High Costs		
Proved reserves	27.0	27.3
Reserve growth	53.0	46.6
Undiscovered	60.0	20.5
Total	140.0	94.4
High Costs with Advanced Technology		
Proved reserves	27.0	27.3
Reserve growth	150.0	82.7
Undiscovered	70.0	30.1
Total	247.0	140.1

[a] Moderate costs assumed by AAPG were $25/barrel in 1986 dollars; moderate costs assumed by ICF Resources were $24/barrel in 1988 dollars. Higher costs were assessed by AAPG at $50/barrel in 1986 dollars; higher costs assumed by ICF Resources were at $40/barrel in 1988 dollars.

[b] Includes indicated reserves, inferred reserves, and mobile and immobile oil recovery.

SOURCE: AAPG (1989a); Kuuskraa et al. (1989).

estimated by AAPG are, however, generally consistent with the estimates announced in 1989 by the USGS and the MMS.

Both the AAPG and the ICF estimates show the strong need for advanced technology to convert the remaining resource base to producible reserves. At moderate cost the volume of resources, exclusive of proved reserves, essentially can be doubled with advanced technology and efficiency. ICF Resources shows an even greater volume of the resource accessible at moderate costs with advanced technology than with high costs alone.

In short, while early U.S. exploration and development could rely on economies of scale, exploiting the remaining resources will generally require economies based on improved efficiencies.

Both AAPG and ICF Resources estimate resources exploitable at moderate cost that are equivalent to three to five decades of current U.S. domestic production rates. At high costs and with advanced technologies, the exploitable resources are obviously greater.

Gas Resources

Estimates of remaining U.S. natural gas resources are regularly made by an industry-based group, the Potential Gas Committee (PGC), and by the DOI (USGS for on land and MMS for offshore; see U.S. DOI, 1989). The most recent comprehensive estimate of the entire natural gas resource base was made by a panel of gas analysts for the DOE in 1988. That panel consisted of representatives from industry, private foundations, state and federal governments, academia, and the estimating agencies and groups. The panel estimated that the volume of natural gas recoverable with existing technology was 1059 Tcf for the continental United States. The panel further reported, by different categories of the resource base, volumes of natural gas that could be exploited at wellhead prices up to $3.00/Mcf and up to $5.00/Mcf (in 1986 dollars) (Table 2-2). While adopting the DOE estimates for existing technology, the AAPG has estimated the volumes of the total resource base exploitable at the above price levels including assumptions of advanced technology. AAPG's estimated volume is equivalent to an 80-year natural gas supply at the current rate of consumption.

ICF Resources, in a report to this committee, made separate estimates for gas accessibility (Table 2-2). With existing technology, ICF Resources estimates volumes that are only marginally higher than the DOE estimates, based largely on reevaluation of potential from nonconventional sources. However, with advanced technology their estimates of moderate-cost natural gas (up to $3/Mcf) from nonconventional sources substantially exceed AAPG estimates; at prices up to $5/Mcf, ICF estimates of some categories of resources are below those of AAPG.

Natural Gas Liquids

Assuming future yields of liquids from conventional natural gas on the order of those historically extracted, remaining volumes of natural gas liquids are estimated to range from about 12 billion bbl at moderate costs to about 20 billion bbl at gas wellhead prices up to $5.00/Mcf. These volumes constitute additions to the liquids potential from remaining crude oil re-

TABLE 2-2 U.S. Natural Gas Resources at Year End 1986 (trillion cubic feet)[a]

	Existing Technology and Efficiency[b]		Advanced Technology and Efficiency[c]	
	<$3.00/Mcf	<$5.00/Mcf	<$3.00/Mcf	<$5.00/Mcf
Proved reserves	166 (162)	166 (187)	166 (163)	179 (187)
Reserve growth[d]	197 (197)	226 (226)	313 (N/A)	483 (N/A)
Undiscovered	144 (144)	233 (233)	202 (N/A)	338 (N/A)
Low permeability	70 (79)	119 (86)	130 (245)	300 (275)
Coalbed methane	8 (13)	12 (26)	40 (51)	90 (65)
Shale	10 (17)	15 (20)	30 (37)	40 (46)
Total	595 (612)	771 (778)	881	1430

[a]Numbers in parentheses are from ICF Resources (Kuuskraa et al., 1989). Note: Mcf = thousand cubic feet.
[b]DOE (1986 dollars).
[c]AAPG (1986 dollars).
[d]Includes inferred reserves, new pool, and reserve growth from gas and gas-associated oil reservoirs.
SOURCE: Kuuskraa et al. (1989); AAPG (1989b).

sources. These estimates may be optimistic because the gas from deeper and unconventional sources is not as rich in liquids.

PRODUCTION TECHNOLOGIES AND PROCESSES

Primary oil recovery depends on the natural energy contained in the reservoir to drive the oil through the complex pore network to producing wells. The driving energy may come from liquid expansion and evolution of gas dissolved in the oil as reservoir pressure is lowered during production, expansion of free gas in a gas "cap," influx of natural water from an aquifer, or combinations of these effects. The recovery efficiency for primary production is generally low when liquid expansion and solution gas evolution are the driving mechanisms. Higher recoveries are associated with reservoirs having water or gas cap drives and from reservoirs where gravity effectively promotes drainage of the oil from the pores. Eventually, the natural drive energy is dissipated. When this occurs, energy must be supplied to the reservoir to produce additional oil.

Secondary oil recovery involves introducing energy into a reservoir by injecting gas or water under pressure. The injected fluids maintain reservoir pressure and displace a portion of the remaining crude oil to production wells. Waterflooding is the principal secondary recovery method and currently accounts for almost half of the U.S. daily oil production. Limited use is made of gas injection because of the gas's value; however, when gravity drainage is effective, pressure maintenance by gas injection can be very efficient. Certain reservoir systems, such as those with very viscous oils and low permeability or geologically complex reservoirs, respond poorly to conventional secondary recovery techniques. In these reservoirs improved geologic understanding and use of enhanced oil recovery (EOR) operations should be employed as early as possible.

Conventional primary and secondary recovery processes, at existing levels of field development, will ultimately produce about one-third of the original oil in place (OOIP) in discovered reservoirs. For individual reservoirs the recoveries range from the extremes of less than 5 percent to as much as 80 percent of the OOIP. The range chiefly reflects the degree of reservoir complexity or heterogeneity. The more complex the reservoir, the lower the achievable recovery.

Of the remaining two-thirds of OOIP in domestic reservoirs (about 340 billion bbl), approximately 30 percent exists as conventionally moveable oil. A portion of this oil can be recovered through advanced secondary recovery techniques involving improved sweep efficiency in poorly swept zones of the reservoirs. For these reservoirs, well placement and completion techniques need to be pursued consistent with the degree of reservoir heterogeneity. Such improved secondary oil recovery can be accomplished using advanced geologic models of complex reservoirs.

The balance of the remaining two-thirds of unrecovered oil is oil that is or will be residual to efficient sweep by secondary recovery processes. Portions of this residual oil can be recovered by tertiary or EOR. The intent of EOR is to increase ultimate oil production beyond that achieved by primary and secondary methods by increasing the volume of rock contacted by the injected fluids (improving the sweep efficiency), reducing the residual oil remaining in the "swept" zones (increasing the displacement efficiency), or by reducing the viscosity of thick oils.

Current EOR technology processes can be broadly grouped into three categories: thermal, miscible, and chemical methods. These processes differ considerably in complexity, the physical mechanisms responsible for oil recovery, and maturity of the technology derived from field applications. Although enhanced oil recovery methods using microorganisms or electrical heating have been proposed, their current state of development is still at the research and initial field pilot stage.

Thermal recovery methods include cyclic steam injection, steamflooding, and in situ combustion. The thermal methods are used to reduce the oil's viscosity and provide pressure so that the oil will flow more easily to the production wells. The steam processes are the most advanced EOR methods in terms of field experience. As a result they have the most certainty in estimating performance, provided that a good reservoir description is available. Steam processes are most often applied in reservoirs containing viscous oils and tars. This is usually done in place of, rather than following, primary or secondary methods. Steam processes have been commercially applied since the early 1960s. In situ combustion, an alternate thermal process, has been field tested under a wide variety of reservoir conditions, but few projects have proved economic and advanced to commercial scale. When oil prices are right, it will probably be applied to moderate-size projects where it is not feasible to use other processes.

Miscible methods use carbon dioxide, nitrogen, or hydrocarbons as miscible solvents to flood the reservoir and can produce 10 to 15 percent of the OOIP. The solvents mix with the oil without an interface and are very effective in displacing the oil from the reservoir. Unfortunately, they do not always achieve a high sweep efficiency. Their greatest potential is enhancing the recovery of low-viscosity oils. Commercial hydrocarbon miscible floods have been operated since the 1950s. Carbon dioxide miscible flooding on a large scale is relatively recent and could make a substantial contribution to EOR production, if prices are right. Only limited field experience with nitrogen is available, but it may be attractive in areas without a ready supply of carbon dioxide and where the reservoir is deep enough to achieve miscibility.

The chemical methods include polymer flooding, surfactant (micellar/ polymer, microemulsion) flooding, and alkaline flooding processes. These methods take advantage of physical attributes of chemicals injected along with a displacing water driver to improve recovery. Polymer flooding is conceptually simple and inexpensive, but it produces only small amounts of incremental oil. It improves waterflooding by using polymers to thicken the water to increase its viscosity to near that of the reservoir oil so that displacement is more uniform and a greater portion of the reservoir is contacted.

Surfactant flooding is complex and requires detailed laboratory testing to support field project design, but it can produce as much as 50 to 60 percent of residual oil. Surfactants are injected into the reservoir to reduce the interfacial tension between the residual oil and flood water to "wash" the oil from the reservoir rock. The surfactant causes the oil droplets to coalesce into an "oil bank" that can be pushed to production wells. Improvements in displacement efficiency clearly have been shown; however, sweep efficiency is a serious issue in applying this method. As demonstrated by field tests, it has the potential to improve the recovery of low- to moderate-viscosity oils. Surfactant flooding is expensive and has been used in few large-scale projects. As a result it is among the least developed of the EOR technologies.

Alkaline flooding is appropriate only for reservoirs containing specific types of crude oils. Surfactants are generated in the reservoir (rather than at the surface) by chemical reactions between injected alkaline chemicals and certain petroleum acids that must be present in the crude oil. Alkaline flooding, although lower in cost and simpler in concept than surfactant flooding, is not well developed. It has been tried only in a few reservoirs and requires considerable development.

UPSTREAM OIL AND GAS ENVIRONMENTAL IMPACTS

Upstream environmental impacts occur in the following activities: exploration, production, and transportation. It is important that activities occur with the least environmental impact that is realistically possible. Only a brief summary of these impacts is provided in this section.

Exploration activities typically involve a variety of superficial surface studies plus shooting a seismic survey. Offshore seismic surveys involve essentially no impact. Land surveys can involve minor disruption of small plants but barely affects animals or structures. Offshore exploration wells are usually drilled from drill ships and have little impact. Land drilling requires clearing roughly 1 acre for rig operations. Drilling mud disposal techniques are well established and not considered hazardous when conducted properly. Blowout accidents are rare, so spillage, explosions, and other related concerns are infrequent. Air pollution is usually negligible.

Typically, production activities involve clearing a drillsite, laying gathering pipelines, and constructing a small processing plant and storage tanks for each lease or unit. Drilling mud and process plant waste disposal are readily manageable and generally considered to have minor impact. When oil and gas are brought to the surface, produced water often accompanies them. It is separated and reinjected into rock at intervals where it either aids production or has no impact on the surface. The wells are carefully cased to avoid affecting other subsurface waters. After a drill rig is removed from a wellsite, the site is restored to original use and only some valves and pipes are visible. Processing plants and tankage are also visible but are relatively small compared to the size of an oil or gas field. Emissions to the air are usually tightly controlled and normally minimal. Accidents are always possible and spills of varying sizes can happen, but good operating procedures minimize such mishaps.

All these activities, either by small or large operators, could have environmental problems associated with them if proper methods and procedures are not followed. However, given correct practices, the probability of permanent environmental damage is small.

Transportation by pipeline, truck, tanker, or barge is uneventful under normal conditions. However, the recent *Exxon Valdez* tanker spill, and the triple accidents the weekend of June 23, 1989, illustrate that accidents that can dispoil waterways and shorelines are possible. Such severe accidents have serious impacts, and efforts must be made to minimize their likelihood.

TIME AND INVESTMENT REQUIRED FOR INCREASED OIL AND GAS PRODUCTION

The time and investment required to bring large amounts of new domestic oil and gas to market are measured in the 5- to 10-year and billion dollar range. Smaller production can be managed on smaller time and investment scales. In this regard new domestic oil and gas production is different from facilities for conversion of nonpetroleum resources into liquid fuels; new oil and gas can quickly be brought to market in small increments at modest costs, while coal and oil shale conversion plants require large multibillion dollar grass-roots plants that do not produce until they are brought into operation a decade or more after planning begins.

To understand the time and investment involved to produce new oil and gas, it is instructive to break the process into the basic steps required. For simplicity the following are defined:

1. *Play/prospect development* consists of those activities required to develop an active exploration interest in a new area. Such activities take 1 to 3 years and may require investments of hundreds of thousands of dollars or more.

2. *Leasing* is that activity required to obtain the right to drill and, if successful, to produce in the area of interest. Normally this takes a few months but can extend to several years in environmentally sensitive areas. Costs can range from tens of thousands to hundreds of millions of dollars depending on the potential target, its apparent attractiveness, and competition for the leases.

3. *Exploratory drilling* results in the first well or wells to determine if hydrocarbons are present. It can take months and cost hundreds of thousands to millions of dollars depending on circumstances.

4. *Delineation drilling* results in subsequent wells aimed at determining the size and character of the deposit. Times and costs are similar to those of exploratory drilling.

5. *Early planning, development drilling, facility construction and initial sales* are those activities required to prepare an Environmental Impact Statement, obtain permits, procure supplies, establish flow, and process and deliver the fluids to an appropriate market. This may require several months to years to accomplish depending on the physical environment. Costs are millions of dollars per 1000 bbls per day of production.

6. *Field development for primary and secondary production* results in refinement of the understanding of the field geology and creation of a reser-

voir management plan to guide subsequent drilling, waterflooding, and eventually EOR, if appropriate.

Once development has begun, field extensions, infill drilling, waterflooding, and EOR generally evolve sequentially. The determinants of action and timing are oil and gas price outlook; reservoir character (amount of remaining hydrocarbons, geological complexity, etc.); and availability of capital and technical personnel. Time and investment required here depend on specific circumstances. For our purposes it can be considered incrementally continuous, until economic incentive dwindles. Ranges for the production of increments from existing fields of 1000 bbl/day of oil are on the order of months and millions of dollars.

Extended reserve growth only requires additional refinement of the reservoir geologic and engineering models through geoscientific and reservoir engineering study, since leases are secured and the presence of hydrocarbons has been confirmed. This is followed by targeted infill drilling with iterative improvement of the models to more completely characterize reservoir heterogeneities and identify hydraulic compartments. This should result in subsequent modification of the planned programs. Since this has not been done on a routine basis, only limited experience is available on the cost and time for target infill drilling programs, but it is apparent that time and investments are relatively modest. Such programs might even be considered a reasonable preliminary stage for implementing an EOR project.

For EOR projects a pilot test is usually required, followed by full field implementation. Depending on the process and the field, preparation for a pilot can take a few to many years and cost on the order of $10 million. Full-field EOR implementation can take 5 to 10 years and millions of dollars per 1000 bbl/day of oil production.

It is essential to note that U.S. industry no longer has the incentive or the ability to aggressively explore and develop new oil and gas resources or to implement EOR on a massive scale. This is because the twin oil price collapses of 1986 and 1988 decimated the domestic industry. Therefore, should the national need develop, a massive scale-up of domestic oil and gas development would probably take 7 to 10 years to accomplish. This estimate is based on the experience following the 1970 oil crisis, when it took about 6 years to double the U.S. oil drilling rig count (see Figure 2-2).

LOSS OF RESERVE GROWTH AND EOR POTENTIAL

As oil fields are depleted, their production rates decrease and operating and maintenance costs escalate. At some point fields reach their economic limit, where a reasonable profit is no longer possible. Clearly, oil fields will reach this point sooner when oil prices are lower, as has been the case in recent years.

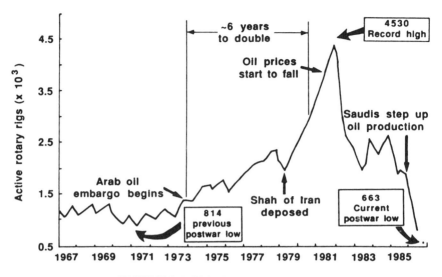

FIGURE 2-2 U.S. rig count since 1967.

When profitable operations are no longer possible, wells are plugged with cement, equipment is removed, surface areas are returned to their preproduction condition, leases are relinquished, and operators abandon the field. Left behind in the reservoir is residual oil that in many cases could be recovered by infill drilling or EOR that is not instituted because economics are unattractive at projected oil prices.

Each year large numbers of oil fields are abandoned (the rate is the highest in the history of the industry) and their target oil is lost for decades, if not forever. This is because the well and equipment investments, the leases, and usually the understanding of the reservoir are lost. Releasing is usually time-consuming and very expensive, if not effectively impossible. It may be impossible because in mature producing areas the mineral interest owners have proliferated over the years, and developing reasonable contracts (mineral leases) is too complex and time-consuming. Under most situations, moving back into an abandoned reservoir will add $4 to $8/ barrel to the cost of the additionally recoverable oil. When this is added to the incremental costs of redevelopment or EOR, it can make the costs prohibitive.

The United States is thus losing potential target oil, mobile unswept, and other oil that could be recovered by EOR at a significant rate. Without higher oil prices or some government action, this potential source of U.S. oil production will be postponed for a long time if not lost forever.

TECHNOLOGICAL OPPORTUNITIES

At the committee's June 1989 workshop (see Appendix B), five experts made presentations on various EOR technologies. Their presentations covered the three dominant process technologies, advanced geoscientific models, and an overview of the status of EOR. Their presentations can be summarized as identifying eight opportunity areas.

1. *Advanced geological reservoir modeling for complex reservoir systems.* Although petroleum has been exploited for decades, detailed understanding of the influence of reservoir flow units and fundamental rock-fluid interactions in complex media is still at a relatively elementary stage. As a result field applications of secondary and tertiary processes often do not perform as expected. The cause of this deviation is often incomplete understanding of the reservoir(s). Recovery, both secondary and EOR, would be greatly improved if advanced geological-reservoir engineering models of the reservoirs could be developed. Improved reservoir characterization would lead to better predictions of where and how remaining oil is distributed, and better methods (using geostatistical techniques along with in situ imaging methods) to characterize heterogeneity at the macroscale and on a scale important for process predictions and would develop understanding and methods for scaling up reservoir description details to the scale of reservoir simulator grid blocks. Targeted infill wells and strategic completions would assure that all hydraulic flow units in a reservoir are contacted and swept. Advanced geological modeling is the essential key to enlarge mobile oil reserve growth.

2. *Extraction technology for immobile oil.* (a) *Advanced miscible flooding* methods have been extensively tested and shown to be effective in well-understood reservoirs. The technology can be characterized as a midlife technology. Current industrial research is primarily focused on optimizing existing technology, with some effort on developing new processes, primarily foams for improving sweep efficiency. Opportunities for future research include improved reservoir characterization; improved process prediction to include quantitative mechanistic descriptions of fingering (unstable intrusion of the displacing fluid into the reservoir oil) and displacement efficiency; more quantitative understanding of heterogeneity effects on miscible processes and the degree of heterogeneity definition required; improved simulation approaches and procedures with innovative computing techniques; and "new" ideas.

(b) *Chemical processes* may be viewed as dramatic extensions to waterflooding, the most common secondary recovery method. Chemicals are added to the drive water to improve sweep or displacement efficiency. Four types of chemical recovery processes can be identified. Permeability modification and profile control uses chemicals to remedy poor sweep efficiency caused by vertical permeability stratification. Several commercial systems are available, and several field tests have been performed. The results are mixed but promising. Predictability is poor. Opportunities for research include improved mathematical models to guide application and process development and development of systems capable of extending treatment away from the well bore deep into the reservoir.

Polymer flooding uses polymers to thicken the drive water so that its viscosity is close to that of the reservoir oil. The process is understood and can be simulated if the reservoir description and rock-fluid interaction properties are known. There is difficulty in propagating the polymer over long distances in the reservoir. The major research opportunities are developing new polymer structures that tolerate high reservoir temperatures, that can move long distances in the reservoir without degradation, and that can be injected at rates higher than those currently used.

Surfactant flooding uses chemicals to improve the displacement efficiency by mobilizing residual oil. The process is understood, and design procedures have been developed. Effective surfactant systems have been developed for moderate-temperature sandstone reservoirs that have relatively low salinity water. Achieving good volumetric sweep is a major concern. Research opportunities include improved surfactants for more hostile reservoir environments and methods to improve volumetric sweep and improved process performance through reservoir characterization.

Alkaline polymer flooding uses the mechanism of in situ formation of surfactants by neutralizing petroleum acids. The process is not well understood, and there have been few successful field tests. The process is only suited to specific reservoirs. Opportunities for research include defining the process mechanisms, developing predictability, and field testing to confirm understanding and demonstrate oil mobilization.

(c) *Thermal methods* are commercially proven processes for enhancing the recovery of heavy oils by reducing their viscosity. Three methods are currently used. Steam stimulation uses a single well for injection, "soak," and subsequent production. It is relatively inexpensive and has rapid response. It results in low ultimate recovery, lost production during "soak" periods, and limited well life and has poor predictability. Opportunities for research include process improvements to increase recovery, improvements in sand control and well materials to improve well life, and greater understanding of the process in order to improve predictability. Steam flooding extends the thermal benefits of steam to greater portions of the reservoir by using separate wells for injection and production. It increases the recovery by increasing the reservoir volume affected by the steam. It also extends the applicability of the method to light oil reservoirs. It suffers from poor sweep efficiency due to steam override and is expensive because of high operating (fuel) costs. In general, predictability is good. Research opportunities include developing methods for improving sweep efficiency both vertically and horizontally.

In situ combustion uses injected air to oxidize part of the oil to raise the temperature of the reservoir. It can extend applicability of thermal methods to thin reservoirs and to greater depths. Unfortunately, the process is difficult to control, has poor vertical sweep, and is expensive, and part of the oil is consumed as fuel. Research opportunities include improving process control, developing methods for improving sweep efficiency, and reducing capital costs. Generally, all of the thermal methods could use techniques for improving sweep efficiency and process control improvements or material improvements to reduce heat losses or the costs of heat generation.

(d) *Advanced EOR methods.* Research on several other EOR methods is under way. These include, but are not limited to, microbial methods and electric heating of the reservoir. These methods provide alternative techniques for achieving the EOR mechanisms, such as reduction in oil viscosity or improvements in displacement or sweep efficiency. Since they are novel, it is difficult to predict whether they will be able to do this in a cost-effective manner. Currently most research is directed toward understanding the mechanisms, with limited field testing under way.

3. *Advanced monitoring.* All recovery processes—primary, secondary, and tertiary—are difficult to monitor. The processes occur away from the well bore, and no techniques are currently available for monitoring their performance at these remote locations. Geophysical methods that could measure properties that allow inferences to be made about the performance of the process would greatly assist control and allow early remedial action to be taken.

4. *More effective production technology.* All production processes require effective well completions. Horizontal drilling offers the opportunity to change the geometry of reservoir flow and increases the contact area between the well bore and the reservoir. Research to determine the influence of well completions on process performance and to develop cost-effective improvements can lead to potential opportunities.

5. *Improved exploration techniques.* Much of the remaining undiscovered U.S. oil and gas resources, particularly on land in the lower 48 states, will be discovered and exploited in small to moderate-size increments. While exploration technology development is highly competitive and proprietary, fundamental research aimed at supporting the development of cost-effective exploration methods for such targets should be a part of DOE's research program. Included could be fundamental geophysical, geological, and geochemical studies. Close coordination with industry in such matters will help ensure maximum benefits and avoid conflicts with proprietary interests.

6. *Technology transfer.* Technology transfer is not a research area but is an essential area to note, particularly for the very important independent producer. In the past, technology was naturally transferred from majors to independents by personnel migration and dissemination from the major companies. This was effective because both independents and majors operated in the same geographic areas. With major producers shifting their investments into frontier areas and international operations, it can be expected that their research will focus more on these areas. There is a substantial body of available technology that independents could use to their benefit if it were available to them. However, it is unlikely, given the lack of technological sophistication and low capitalization of independents, that this technology will be exploited by them in the normal course of events. Thus, a public sector program of technology transfer could help assure that unrecovered domestic oil resources are conserved and efficiently developed. The development of such a program is an important challenge to the government, but was not considered in any detail in this study.

DOE RESEARCH PROGRAM

With the exception of exploration methods, DOE currently supports some research in all of the areas of oil and gas research where this committee sees technological opportunities. From a survey by the National Petroleum Council (NPC), corporate research likewise addresses these areas to varying degrees (NPC, 1988). However, the total expenditure, in oil and gas recovery research, public and private, is small relative to the value of oil and gas produced or to the value of oil and gas reserves added through improved recovery. According to the NPC report, industry expenditures for recovery research on reservoir characterization and EOR processs technology improvement averaged about $195 million yearly for the first half of the 1980s. (The Energy Information Administration indicates much larger expenditures, i.e., $645 to $801 million per year in the mid-80s, but their category of oil and gas recovery is broader and includes all upstream R&D activity [EIA, 1987a]). Public federal and state expenditures for such research is on the order of $40 million annually. The current DOE expenditures of about $24 million per year for enhanced oil recovery are substantially less than for DOE's expenditure of about $86 million per year for other liquid fuels (this does not count the large budget for other uses of coal). Considering the contribution of domestic oil and gas to U.S. supply and the potential of these resources to add quickly to future supplies, public research efforts should be brought into balance with expenditures for other resources, and sufficient incentives should be provided to enlarge corporate research on domestic oil and gas recovery.

SUMMARY

This chapter has shown that in the price range of $25 to $50/barrel for oil, advanced oil recovery technologies could permit the United States to maintain a significant level of domestic oil production for several decades. This would allow time for R&D on converting nonpetroleum resources into liquid fuels. The next chapter addresses the economics of various conversion technologies producing liquid transportation fuels from heavy oil, tar sands, natural gas, and nonpetroleum resources.

3

Production Costs for Alternative Liquid Fuels Sources

In this chapter the costs of several fuel production technologies are estimated to help compare the economic attractiveness of different feedstock, process, and fuel combinations. This chapter also considers the conversion of tar sands, oil shale, coal, natural gas, and wood into transportation fuels as well as the production of ethanol from corn and the use of compressed natural gas (CNG) in vehicles. Cost estimates for electric and hydrogen-powered vehicles are not included. Economic estimates are based only on U.S. resources, but, since there is currently so much interest in methanol, some analyses address methanol produced abroad using natural gas supplies less costly than U.S. natural gas.

Using economic assumptions specified by the committee, literature-estimated economic parameters of the processes were initially compiled by Bernard Schulman and Frank Biasca of SFA Pacific, Inc. (Schulman and Biasca, 1989). These estimates were further refined and updated by various committee members. The final results were used to estimate production costs, on a consistent basis across all feedstocks and technologies, based on current technical understanding. However, many of the technologies have not been commercially implemented, and there remains a high degree of uncertainty about many of the cost elements as well as the total costs.

STRUCTURE OF THE ANALYSIS

Cost analysis of transportation fuels manufactured from alternative energy resources relies on estimates of resource costs, technological assessments of the specific production processes, and assumptions about the economic environment. Together these considerations form the basis for estimating the total cost of producing and using the alternatives. Costs are expressed in per barrel crude oil equivalent that would make spark ignition vehicle fuel from the alternative energy resource just as expensive to the end user as gasoline from crude oil (Appendix D).

Technological assessments for each process take account of the types and quantities of feedstocks utilized, the capacity and investment cost of a typical facility, nonenergy and energy operating and maintenance (O&M) costs, and the value of by-products produced. For fuels characterized by an energy density different from that of gasoline, a gasoline equivalency factor is used. Also, any additional capital costs for alternative-fueled automobiles are included in the analysis.

Economic assumptions include prices of the various energy and nonenergy feedstocks, prices of energy used in plant operation, the real discount rate and the corresponding annual capital charge factor, costs of crude oil refining, and costs of fuel distribution and marketing. Where possible, energy costs are expressed as functions of the crude oil price. Nonenergy costs are estimated independently of oil prices, even though the committee recognizes that real (inflation-adjusted) construction costs may vary with crude oil prices and with the development rates of alternative energy facilities. The analysis assumes a gradual growth of the industry, not a crash program under which construction costs could rise rapidly.

Two different real (inflation-adjusted) after-tax annual discount rates within a discounted cash flow (DCF) framework are considered—10 and 15 percent—with corresponding annual capital charge factors of 16 and 24 percent. The 10 percent discount rate is based on average historical returns required for equity capital in financial markets and on average historical returns earned by physical capital in U.S. industry. The 15 percent discount rate is based on estimates of typical hurdle rates (minimum estimated rates of return) required by corporations for investments or on costs of capital for risky projects, although hurdle rates might be even higher for particularly risky projects.

For each technology the cost per equivalent oil barrel, in 1988 dollars, is calculated as the summary cost measure, based on the assumed oil price environment. This calculation begins with the crude oil price, here defined as the average price of crude oil imported into the United States. Prices of natural gas, electricity, and corn (as a feedstock) are calculated based on the crude oil price (see Appendix D).

It is assumed that conversion facilities for coal are located on the Gulf Coast and that facilities for natural gas conversion, underground coal gasification, oil shale mining, ethanol production, and tar sands pyrolysis or extraction are located at the resource site. Natural gas compression would occur at the point of end use. Thus, natural gas used as a feedstock is priced at an estimated wellhead cost, and natural gas used in plant operations or compressed into CNG is priced at a delivered cost to the industrial sector. Coal price includes a transportation charge to the Gulf Coast in addition to the mine-mouth price. No royalties for oil shale or tar sands are included in the costs, because (1) the magnitude of royalties would vary with the profitability of the resource and conversion technology combination and (2) royalties would be driven toward zero as the overall profitability were itself driven toward zero.

Per-gallon product costs are calculated by adding together feedstock costs, energy and nonenergy O&M costs, and annual capital cost and subtracting by-product credits, all on a per-gallon basis. For processes that produce a product similar to crude oil, cost per equivalent oil barrel is reported as equal to the product cost per barrel. No further calculations are required.

For processes that produce gasoline directly, a gasoline substitute, or products intermediate between gasoline and crude oil, gasoline equivalent costs are first calculated. Cost per equivalent oil barrel is then estimated by adjusting gasoline equivalent cost. To calculate gasoline equivalent costs, product cost per barrel is multiplied by the gasoline equivalency factor.

The gasoline equivalency factor has a value of 1.8 for methanol, reflecting a 10 to 18 percent efficiency gain for automobiles. A value of 1.5 was assumed for ethanol, but its octane advantage and potential for enhanced efficiency could lead to lower ratios. For CNG the analysis is conducted in terms of output per equivalent gallon of gasoline, so that the gasoline equivalency factor is 1.0 (although optimized CNG vehicles may have efficiencies greater than gasoline engines). For the Shell middle distillate synthesis (MDS) process, the tar sands solvent extraction process, and direct coal liquefaction, a gasoline equivalency factor of 1.0 is used, reflecting the opportunity to use the outputs from these processes in the production of gasoline.

Three adjustments to the gasoline equivalent cost are made to obtain cost per equivalent oil barrel. The distribution and marketing costs of the product, net of corresponding costs for gasoline, are added to the gasoline equivalent cost. A refining credit, based on the historic relation between prices of gasoline and crude oil, is subtracted for those products that produce gasoline or a direct substitute for gasoline (methanol, ethanol, and gasoline through the methanol-to-gasoline [MTG] process). A smaller spread is subtracted for those products intermediate between crude oil and gasoline. Finally, when appropriate, the additional annualized incremental costs for methanol- and CNG-fueled vehicles (above the costs of gasoline-fueled vehicles) are added.

The procedure outlined above provides an estimate of the cost per equivalent crude oil barrel based on a given crude oil price. Two different modes of analysis are used to select the crude oil price, referred to as *endogenous price determination* and *exogenous price determination*.

For exogenous price determination a particular crude oil price is assumed

PRODUCTION COSTS 43

as part of the scenario specification (see Chapter 1). Two of these scenarios, assuming a $20/barrel and a $40/barrel crude oil price, have been the basis for complete cost calculations. Using this procedure, if the cost per equivalent crude oil barrel exceeds the crude oil price, the difference between these figures can be interpreted as either the amount of subsidy or the degree of economic improvement required to make that process and fuel competitive with gasoline from crude oil. If the difference is negative, it can be interpreted as the rents or profits (above the discount rate or cost of capital) that could be obtained by using the alternative energy resources to supply fuel.

For endogenous price determination, crude oil price is not an input to the analysis but is calculated as the price that would make the specific process and fuel just competitive. More precisely, the crude oil price is chosen such that the calculated cost per equivalent oil barrel is exactly equal to the crude oil price: No profits would be earned above the normal cost of capital or discount rate. Using this procedure, electricity, natural gas, and corn prices are also chosen to be consistent with the crude oil price and the cost of the specific product. The cost per equivalent crude oil barrel can then be interpreted as the crude oil price at which the fuel (based on a given resource and process) would be just competitive (without a subsidy) with gasoline from crude oil. If crude oil prices are in fact higher than the cost per equivalent barrel, rents or profits (above the discount rate or cost of capital) would be available, while for lower prices the alternative process would not be commercially viable without a subsidy.

COST ESTIMATES FOR THE VARIOUS TECHNOLOGIES

Based on the procedure outlined above, cost estimates have been developed for each technology using domestic feedstocks (see Appendix D). This section presents the basic results.

Figure 3-1 shows the costs of various alternative fuels, based on a 10 percent discount rate and endogenous determination of energy prices. Current commercial technologies include natural gas used as feedstock to produce methanol (NG > Methanol), natural gas converted through a methanol-to-gasoline process (NG, MTG), compressed natural gas (Compressed NG) used directly in automobiles, and corn distilled into ethanol (Corn > Ethanol). Technologies successfully demonstrated on a commercial scale but not yet commercialized include coal used as a methanol feedstock (Coal > Methanol), wood used as a methanol feedstock (Wood > Methanol), and indirect coal liquefaction in a methanol-to-gasoline process (Coal, MTG). There is still considerable uncertainty regarding the economically accessible biomass resource base, and research is on-going to overcome problems in conversion processes and improve the economics of producing liq-

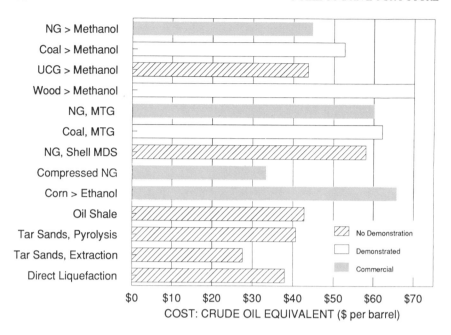

FIGURE 3-1 Estimated costs of alternative fuels at 10 percent DCF with endogenous calculation of energy prices (see text for abbreviations and further detail). New estimated reduced capital and operating expenses for entrained-flow coal gasification could lead to coal-to-methanol costs of about $40/barrel.

uid fuels. Technologies that have not been successfully demonstrated on a commercial scale include underground coal gasification producing methanol (UCG > Methanol), the Shell Middle Distillate Synthesis process (Shell MDS), pyrolysis of oil shale (Oil Shale), pyrolysis of tar sands (Tar Sands, Pyrolysis), solvent extraction of tar sands (Tar Sands, Extraction), and direct liquefaction of coal (Direct Liquefaction) (see Appendix D, Table D-9).

At a 10 percent discount rate, estimated costs range from $25/barrel to above $70/barrel, with most exceeding $40/barrel. Only compressed natural gas, direct coal liquefaction, and tar sands (solvent extraction) have estimated costs below $40/barrel. Methanol produced from domestic natural gas or underground coal gasification, hydrocarbon fuels from oil shale conversion, and pyrolysis of tar sands have estimated costs between $40 and $50/barrel. These processes could become economically viable with significant technological advances and a high world oil price. All other technologies have oil equivalent costs exceeding $50/barrel and will be less economical unless significant cost reductions are realized.

Production of methanol from coal and wood is of special interest since both raw materials are potentially major domestic methanol sources. The endogenous energy prices for coal and wood were estimated to be $53/barrel and $70/barrel, respectively, with corresponding feedstock costs of $16/barrel and $24/barrel (Table D-3). Much of the difference in cost is attributable to higher investment and operating costs related to the small assumed scale of the wood-based process. Increasing the scale of the wood-based process would narrow or eliminate the difference between nonfeedstock costs for the two processes. Technological advances can generally be applied to plants using either feed. The synthesis costs are independent of source of the carbon monoxide-hydrogen mixture used and cost reductions would also be applicable to natural gas-based plants. Advances in the gasification section are generally applicable to both coal and wood. While advances in technology can reduce methanol costs from wood and coal materials, the substantially lower projected cost of methanol from low-cost natural gas indicates that as long as overseas natural gas prices are low, importation will be the lowest cost methanol source. Synthesis gas-based methanol and hydrocarbons, because of energy losses in both gasification and synthesis, are expected to remain a more expensive route to transportation fuels than direct liquefaction or manufacture from high-grade oil shale.

As discussed later (see Figure 3-5), the relative cost of methanol from natural gas is highly sensitive to the cost of natural gas (see Appendix D). The lower estimated cost for direct coal liquefaction versus oil shale reflects the progress made over the past decade in reducing coal liquefaction costs and the lack of published advances in oil shale conversion technologies.

Major costs incurred in producing fuel from alternative energy resources include those for capital, feedstock, and O&M (Figure 3-2). The incremental cost of purchasing a methanol- or CNG-fueled vehicle, spread over the life of the vehicle, is also shown as a positive component of the oil equivalent cost. For most of the technologies, feedstock and capital costs are the two largest cost categories. By-product credits are small except for the production of ethanol from corn. Ethanol production results in large quantities of organic by-products that the present analysis assumes can be sold for one-half the value of the initial feedstock. However, the market and price at which they can be sold is highly uncertain. Laboratory results for producing ethanol from other forms of biomass such as wood, through advances in biotechnology, show promise but are not included in the present analysis.

Several negative components appear in Figure 3-2. The sum of positive components minus the sum of the negative components is consistent with the cost estimates in Figure 3-1. By-product credits are shown as negative components, since they would subtract from the net cost of producing fuel.

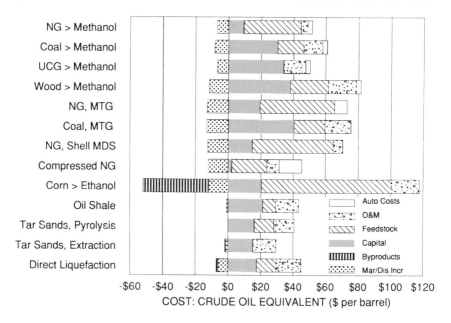

FIGURE 3-2 Components of total cost for alternative fuels at 10 percent discounted cash flow with endogenous price calculation. (O&M, operation and maintenance costs; Mar/Dis Incr, increment associated with marketing, refining, and distribution.)

Two components—one negative and one positive—have been aggregated into a component covering refining, marketing, and distribution. The negative component is the historical spread between refined gasoline price and the crude oil price. The positive cost component is the cost of marketing and distributing ethanol and methanol, net of the equivalent cost of marketing and distributing gasoline. The sum of these two components is negative for most of the processes and thus appears in Figure 3-2 as negative cost components.

Figure 3-3 provides cost data similar to the data in Figure 3-1 for both the 10 and 15 percent real discount rates. This diagram shows that the costs of these capital-intensive technologies are very sensitive to the cost of capital. The greatest sensitivity occurs for those technologies that require the largest per-barrel investment cost (see Appendix D, Figure D-1).

Cost estimates for two crude oil prices, $20/barrel and $40/barrel, indicate that crude oil equivalent costs of the various technologies do depend on crude oil price but that the sensitivity is relatively small (Figure 3-4). Two factors explain the impact of crude oil price: the cost of energy inputs and the marketing and refining credit for fuels that directly lead to gasoline

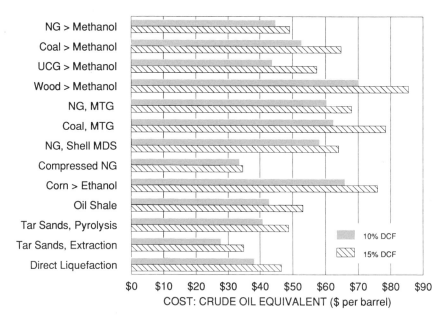

FIGURE 3-3 Role of discount rate on cost for endogenous energy prices.

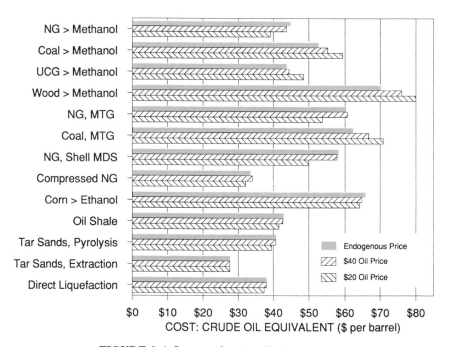

FIGURE 3-4 Impact of crude oil price on cost.

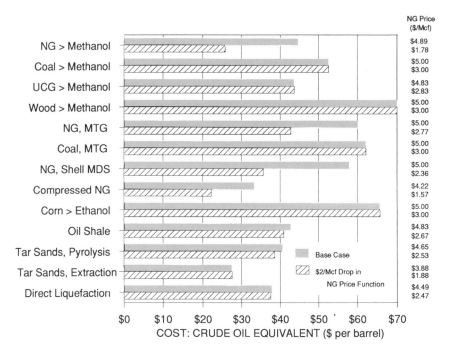

FIGURE 3-5 Impact of natural gas price on cost.

or a gasoline substitute. Higher oil prices result in a greater credit. The net effect of these two factors will determine the overall oil equivalent cost.

Sensitivity of costs to natural gas prices were calculated by reducing the natural gas price by $2/million British thermal units at each oil price (Figure 3-5). While this variation in the natural gas price function reduced the estimated cost for most technologies, it led to far less cost variation than did the cost of capital variations, except for technologies using natural gas as feedstock.

The importance of capital and feedstock costs is indicated in Figure 3-6. Capital plus feedstock costs exceed $50 per oil equivalent barrel for all MTG processes, for ethanol from corn, for methanol from wood, and for the Shell MDS process. Two methanol-producing technologies—using natural gas and coal as feedstocks—have feedstock plus capital costs exceeding $40/barrel. Unless R&D advances lead to great reductions in investment costs or to great increases in feedstock conversion efficiency, such technologies are unlikely to become economical. Some current research efforts suggest that capital costs for wood to methanol conversion could be reduced significantly.

Sensitivity tests for domestically produced methanol (using natural gas as a feedstock) illustrate major uncertainties about the costs of this gasoline substitute. Ranges of crude oil equivalent cost estimates are based on sensitivity tests for five parameters, varied one at a time, using the endogenous price determination method. Results are presented in Figure 3-7 and in Table D-10 (Appendix D).

The base case assumes that automobiles fueled with methanol will enjoy a 10 to 18 percent efficiency gain over gasoline-powered vehicles. This assumption translates to a gasoline equivalency factor of 1.8. Methanol vehicle efficiencies that are 15 percent greater than in the base case (bar labeled "15% More") and 15 percent smaller (bar labeled "None") are illustrated by the pair of bars in Figure 3-7 denoted by "Automobile Efficiency Gain." The former is based on a gasoline equivalency factor of 1.57, the latter on a gasoline equivalency factor of 2.06. This range of automobile efficiency variations leads to a $14/barrel variation in crude oil equivalent cost of methanol.

The investment cost of a methanol production facility is increased by 25 percent and decreased by 20 percent for the third set of sensitivity tests. This range of investment costs leads to a $5/barrel variation in the oil equivalent cost of methanol.

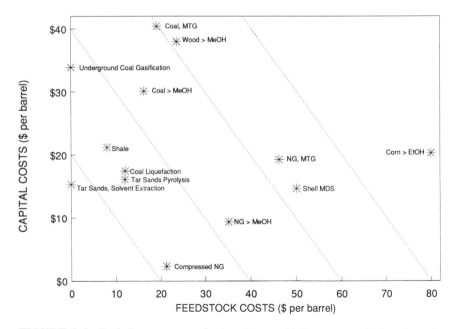

FIGURE 3-6 Capital costs versus feedstock costs (dollars per equivalent barrel). Dotted diagonal lines show combinations of these two costs totaling to $20/barrel, $40/barrel, $60/barrel, and $80/barrel. For shale, mining costs are included in the feedstock costs, whereas for tar sands they are not.

The real discount rate is varied from a low of 5 percent to a high of 15 percent for the fourth set of sensitivity tests. This range leads to a $9/barrel variation in the cost of methanol.

Finally, the differential between the cost of a methanol-fueled automobile and that of an equivalent gasoline-powered vehicle is reduced to zero from the base case of $200/car and increased to $500/car. This range leads to an $8/barrel variation in the crude oil equivalent cost of methanol.

Sensitivity tests for methanol production from natural gas show why methanol is more likely to be produced abroad than in the United States. These tests varied a hypothetical natural gas feedstock price from a high of $3/Mcf to a low of $1/Mcf. Three different investment cost conditions were also compared: estimated investment cost in the United States, that cost increased by 25 percent, and that cost increased by 75 percent, to evaluate possibly higher investment costs in remote foreign locations. O&M costs were increased along with investment costs. Finally, costs were added for transportation of methanol to the United States. Methanol costs for different assumptions are shown in Figures 3-8 and 3-9. Endogenous crude oil price is assumed, but natural gas price is varied independently to assess sensitivity specifically to the price of gas, even though natural gas and oil prices are likely to be linked.

The cost estimates at the far left in Figures 3-8 and 3-9 are based on U.S.

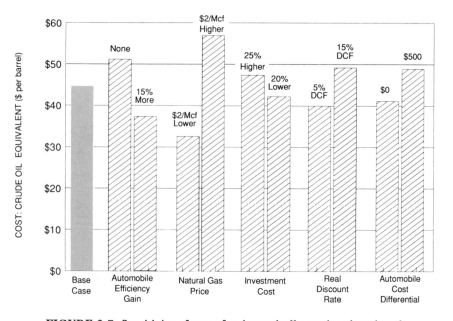

FIGURE 3-7 Sensitivity of costs for domestically produced methanol.

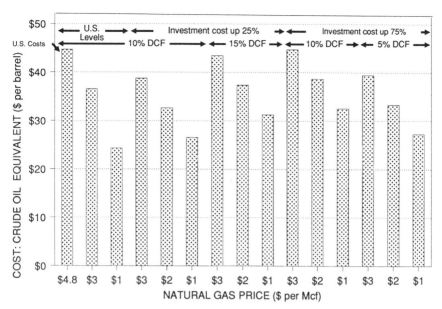

FIGURE 3-8 Total cost for methanol production (in dollars per barrel at 1988 prices). See Figure 3-9 for cost components.

conditions, including a natural gas price of $4.80/Mcf (this cost was determined as the endogenous natural gas price). The next two estimates are also based on U.S. investment and O&M costs, but natural gas price is set at $3/Mcf and $1/Mcf, and a $4/barrel cost is added for transportation of the methanol to the United States. These figures show the sensitivity of the total cost to natural gas prices. The remaining cost estimates represent conditions that might characterize natural gas-to-methanol production in remote locations. These 12 estimates are separated into four groups of three cases. Cost changes are based on estimates developed by Bechtel, Inc. (California Fuel Methanol Study, 1989).

Two groups represent investment costs 25 percent above U.S. levels and corresponding increases in O&M costs. These values might characterize methanol production in the Middle East. Discount rates of 10 and 15 percent are presented for these two groups. For these two groups crude oil equivalent costs range from $26/barrel, were natural gas priced as low as $1/Mcf and the real discount rate at 10 percent, to $43/barrel for a natural gas price of $3/Mcf and discount rate of capital of 15 percent.

The final two groups represent investment costs 75 percent above U.S. levels and 64 percent increases in O&M costs. These values might characterize methanol production in northwestern Australia. A very low discount

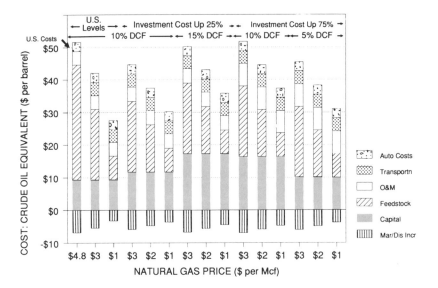

FIGURE 3-9 Components of total cost for methanol production (in dollars per barrel at 1988 prices). (The refining, marketing, and distribution increment—Mar/Dis Incr—should be subracted from the sum of the other costs to obtain the crude oil equivalent costs.)

rate of 5 percent is shown in addition to that of a value of 10 percent to illustrate values that might be obtained if, for example, the Australian government promoted development through low-interest loans or loan guarantees. For these two groups crude oil equivalent costs range from $27/barrel, were natural gas priced as low as $1/Mcf and the cost of capital subsidized (5 percent), to $44/barrel for a natural gas price of $3/Mcf and discount rate of 10 percent.

Estimates for methanol production in remote locations vary greatly from estimates for the continental United States. Investment cost, natural gas feedstock cost, and the cost of capital can each greatly influence cost estimates. The cost of transportation for the finished methanol is a smaller source of variation. Production at different locations is likely to be characterized by quite different values of the major cost determinants.

U.S. production costs for methanol would be significantly higher than in foreign locations characterized by low natural gas prices and modestly increased capital costs. This result strongly suggests that methanol produced in the United States is unlikely to be economical compared with methanol from remote foreign locations. To the extent that methanol is used in the United States, then, it will likely come from remote foreign locations. The committee's cost estimates, based on continental U.S. conditions, should therefore not be used to assess the probable costs of methanol to be used in the United States.

ISSUES OF FUEL DISTRIBUTION AND USE

The previous sections have presented cost estimates of producing and using transportation fuels manufactured from alternative energy resources. This section further addresses issues associated directly with the distribution and end use of such fuels. Changing the fuel infrastructure and vehicle ownership patterns for alternative fuel use is a truly formidable barrier. Environmental issues are addressed in Chapter 5.

On the other hand, many of the processes provide a product similar to crude oil. Such products would be refined into gasoline and other petroleum products that would be used just as fuels refined from conventional crude oil. These products (as well as reformulated gasolines) would not require important changes to the distribution system or to automobiles and are not discussed further in what follows.

End-Use Issues

Alcohol fuels such as ethanol and methanol would be used in automobiles in much the same way as conventional fuels. Although alcohols weaken and corrode some automobile materials, automotive redesign would solve any problems.

Natural gas can be used directly as fuel in an engine. Because of its low liquefaction temperature, natural gas would normally be stored in the gaseous state as CNG at about 3000 psi. Although engineering advances might be expected before large-scale introduction of CNG vehicles into the United States, the basic technology is already commercially available: Several countries are already using CNG vehicles to a limited extent.

In addition to single-fuel vehicles, multifuel engines that can use any combination of gasoline, methanol, and ethanol are under development by automobile companies. These vehicles would allow the use of either fuel during a potential transition from a gasoline to a methanol fuel system. While such engines have important flexibility advantages, they could not be optimized with respect to both fuels. Therefore, they cannot be expected to take advantage of the special properties of some fuels, such as the high octane of methanol. Multifuel vehicles can also be designed to use either natural gas or gasoline.

Both single-fuel and multifuel vehicles using alcohols or CNG will be more costly than vehicles using gasoline engines. Once large production volumes are obtained, it is estimated that multifuel vehicles operating on methanol and gasoline will cost at least $100 to $300 more than conventional gasoline vehicles; these estimates are also a best guess for what a dedicated methanol vehicle might cost. Multifuel autos operating on CNG and gasoline will cost about $700 to $1000 more (DeLuchi et al., 1988b).

Additional purchase costs of multifuel engines have been incorporated above in the cost estimates for methanol ($200) and for CNG ($1000). The additional purchase cost was translated to an equivalent per-mile additional operating cost. This per-mile equivalent operating cost was derived so as to give the same discounted present value of costs to the consumer as would the one-time additional automobile purchase cost. (See Appendix D.)

Vehicle performance could vary among the alternative fuels. Consumer acceptance of these fuels can be expected in turn to depend on vehicle performance. The road performance—acceleration, hill-climbing ability, and so on—of CNG- and alcohol-fueled vehicles is comparable to that of similar gasoline-fueled vehicles. For optimized engines the road performance of alcohol-fueled vehicles can be expected to be slightly better than their gasoline-fueled equivalents.

Methanol and CNG have lower volumetric energy densities than gasoline. Unless methanol and CNG vehicles carry more fuel, a shorter driving range between refuelings would be inevitable. Increasing the size of the fuel tank so as to increase the range may encroach on passenger space, especially in small cars. Compensating increases in automobile size would decrease fuel economy. It remains to be seen how the automobile industry would adjust to this challenge.

Methanol's relatively low vapor pressure and high heat of vaporization hinder satisfactory cold start and driveway performance. Fuel vapor pressure determines the air-to-fuel ratio delivered to the engine. Vapor pressure decreases as its temperature decreases. Thus, the ratio of air to vaporized fuel of methanol is too high for satisfactory cold-start ignition. However, a small amount of gasoline added to methanol (around 15 percent; this is known as M85 fuel) seems to reduce the cold-starting problem—at least in moderate climates—without exacerbating water miscibility problems in the fuel tank. Cold starting is not a problem with CNG vehicles.

The octane number of natural gas and methanol is higher than for gasoline, permitting optimized engines for these fuels to have higher compression ratios and better thermodynamic efficiency. Multifuel engines, however, need to be operable on the lower-octane gasoline. These multifuel vehicles would not gain the fuel-economy benefits of these alternate fuels. In the cost analysis above, a range of fuel-economy assumptions in comparison to gasoline-fueled vehicles was used.

Increased wear has been reported in engines using methanol, but it appears that this problem can be solved with appropriate reformulated motor oils. These oils may be more expensive and may require more frequent oil changes. Natural gas engines have durability equal to or greater than gasoline engines. Thus, it appears that from an engine durability standpoint there is no insurmountable problem with these alternative fuels.

Fuel Distribution Issues

Bulk distribution of alcohols would create additional problems, but problems that could be overcome. The bulk distribution system for liquid fuels depends primarily on pipelines and secondarily on river and ocean barges from U.S. refineries, with terminals located near major centers of demand. Trucks are used to deliver fuels to retail service stations. Because present fuels are not miscible with water, the current system is not kept water-free. Water-free facilities would be required for alcohol-gasoline blends because of phase separation problems (pure alcohols would be less of a problem), and the use of methanol would require changes in materials. Because of the solvent properties of methanol, pipelines and tankage must be cleaned before methanol is introduced into facilities that have been used for petroleum products. Distribution of methanol by barge, rail car, and tank truck, rather than pipeline, is quite feasible at similar or increased cost per gallon.

Natural gas distribution for CNG may not be a problem because there is currently excess natural gas pipeline capacity. However, that capacity might be insufficient in some regions if natural gas became a major transportation fuel. But pipeline capacities could be expanded in adequate time. Since many homes already have gas delivered directly to them, home-fueling stations for natural gas for overnight refilling are possible. However, this option would entail the high cost of a compressor and may involve safety risks.

Retail distribution of alternative fuels would require start-up investments to modify the existing refueling system. For example, the cost of a refueling station for CNG vehicles would be about $300,000 (Sperling and DeLuchi, 1989). These additional costs, to establish a CNG refueling station, comprise the CNG capital costs underlying the economic analysis presented earlier in this chapter. Methanol refueling stations may require an additional cost of around $40,000, perhaps more depending on the need for a new vapor control system.

Potential purchasers of CNG- or methanol-fueled vehicles would be unwilling to buy unless a refueling system were in place. And there would be little incentive to establish a refueling system unless there were customers expected to purchase the fuel. Multifuel vehicles represent a natural transition response to this "chicken and egg" problem. Additional early steps toward problem solution might involve facilities dedicated for vehicle fleets, both public and private. The U.S. light-duty vehicle fleet market is estimated at 0.65 million to 1.3 million bbl/day (10 billion to 20 billion gal/year). Although not all of the fleet market is suitable for alternative fuels, the market could well be large enough to provide the beginnings of a CNG or methanol refueling system. Public "filling stations" may remain rare until there are many vehicles, although fuel companies in California are now establishing methanol fuel outlets for demonstration fleets.

In summary, with additional vehicle development work and demonstration programs, alternative fuels could be used successfully in current transportation engines. With the possible exception of reduced driving range, performance should be acceptable for both CNG and M85. Broad use of either fuel involves significant issues of bulk and retail distribution. However, these are cost, not technical, issues that could be solved if alcohol or CNG were economically competitive with conventional or reformulated gasoline.

CONCLUSIONS

The results of the preceding analysis for producing liquid fuels from alternative energy resources suggest that none of these alternatives can be expected to provide transportation liquids at as low a cost as is currently possible by refining crude oil. Even if crude oil prices were to increase by 50 percent above current levels, only a few of these options would be economically viable.

Economic acceptability of the various processes would come about if there were significant increases in the world crude oil price, the U.S. energy policy environment, the technical characteristics of these processes, or a better understanding of their economics.

If world oil prices were to increase above $30/barrel and were to stay at that elevated level, some processes would become economically attractive under current cost estimates. More would become attractive if oil prices were to increase above $40 to $50/barrel range and remain there for extended periods.

Likewise, U.S. energy policy could be altered so as to apply a large premium, say $10/barrel, to energy conservation or production activities that reduced U.S. petroleum imports. Under such an aggressive policy some processes would become viable immediately in some situations.

Finally, R&D activities might decrease the production costs to below those estimated in this chapter. In particular, technological improvements might reduce investment costs or increase the conversion efficiencies of the various processes (see Figure 3-6 for the magnitudes of these two costs). Such changes could significantly reduce the overall costs such that the various technologies could become economically viable at lower world oil prices or with less aggressive energy policy stances than suggested above.

4

Conversion Technologies and R&D Opportunities

Various technologies can be used to convert heavy oil, tar sands, oil shale, coal, biomass, and natural gas into liquid transportation fuels. These technologies, opportunities to reduce their costs, fuel properties, environmental aspects, and R&D recommendations for the U.S. Department of Energy (DOE) are discussed in this chapter. More information on some of the technologies can be found in the appendixes. The technologies are addressed to the extent that they are major areas for DOE: some sections are kept brief because of their lower relevance (see Chapter 6).

The hydrogen-to-carbon (H/C) ratio of these resources must be adjusted to that of transportation fuels; pyrolytic processes remove carbon, and hydroprocessing adds hydrogen. This adjustment is a major expense and consumer of energy in these processes. In methanol production and indirect liquefaction (Fischer-Tropsch [F-T] and methanol-to-gasoline [MTG]) processes, the entire feedstock is first converted to synthesis gas (mixture of hydrogen and carbon monoxide)—a major cost and energy consumption step. The next section addresses R&D for cost reduction of hydrogen and synthesis gas. The following sections address the individual conversion technologies.

PRODUCTION OF HYDROGEN AND SYNTHESIS GAS

Production of hydrogen and synthesis gas (syngas) is central to converting fossil resources into transportation fuels. They are required to adjust the H/C ratio of these resources and to remove undesirable elements (inorganic, N, S, and O). Conventional transportation fuels have an H/C ratio of approximately 2 (methanol has a ratio of 4; however, from an energy viewpoint it can be considered approximately a mixture of CH_2, the fuel, and

H_2O). Coal clearly requires the largest adjustment in the H/C ratio, but even tars and higher-boiling petroleum fractions require major adjustment (Figure 4-1). The same considerations apply to the manufacture of methanol or hydrocarbon fuels from biomass.

Conversion of natural gas (CH_4) to liquid transportation fuels, in contrast, requires reduction of the H/C ratio. The conventional pathway, as in methanol manufacturing, is to convert the methane to syngas, the first step in hydrogen manufacturing. This syngas can then be converted directly to methanol or to hydrocarbon fuel or it can be processed further to produce hydrogen for hydroprocessing. Production by steam reforming or production by partial oxidation are the standard processes for syngas and hydrogen manufacture.

These natural gas conversion processes are increasingly used in refining and chemical manufacturing. The ample world supply of low-cost natural gas indicates a continuing international engineering and R&D effort aimed at efficiency improvement and cost reduction of these processes. While there will be a continuing need for improved high-temperature materials and catalysts, there is little need for a U.S. government-supported effort on the high-temperature conversion section of these methane-based processes.

When international crude oil prices are greater than about $30/barrel, the price of domestic natural gas is predicted to exceed $4/thousand cubic feet (see Appendix D, Table D-1) and production of hydrogen and syngas from coal becomes competitive. Since this is also the range of equivalent crude prices where use of coal liquefaction and shale oil are competitive, attention to reducing the cost of production of hydrogen and syngas from coal is important.

Table 4-1 shows the amount of hydrogen consumed for conversion of Illinois No. 6 coal to a syncrude suitable for further refining to transportation fuels. Hydrogen used for heteroatom removal is about the same or greater than that added to the liquid product, and a larger fraction of consumed hydrogen is found in C_1-C_3 gases. The total corresponds to 142 to 226 m³/barrel (5000 to 8000 ft³/barrel) of syncrude, and an additional amount would be consumed in refining to transportation fuels. An eventual commercial technology might achieve a 50 percent reduction in hydrogen losses to the by-products.

Equipment for manufacturing this large amount of hydrogen from coal requires a major fraction of the capital investment—approximately 25 percent and a similar fraction of total coal consumption.

In coal gasification water is the primary source of hydrogen through its reaction with carbon to form carbon monoxide. This reaction is highly endothermic, and a large amount of heat is supplied by burning coal with oxygen. This can be done in a single reactor by using a mixture of oxygen and steam or by circulation of hot solids or gas plus steam through a reac-

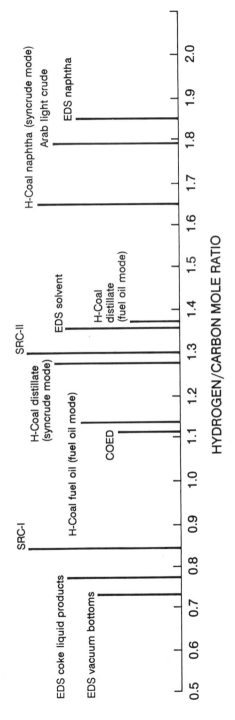

FIGURE 4-1 Hydrogen-to-carbon ratios for various hydrocarbon sources and end products. SOURCE: Whitehurst (1978)

TABLE 4-1 Hydrogen Sources and Uses in Direct Coal Liquefaction

	Pound/ 100 lb Coal	Percent of Total
Hydrogen Sources		
From coal fed	5.33	45
From H_2 gas added	6.39	55
Hydrogen in Products		
In distillable liquids	8.23	70
In by-products		
C_1-C_3 hydrocarbon gases	1.26	11
Heteroatom gases	1.24	11
Resid and unconverted coal	0.99	8

NOTE: Data for Wilsonville Run 257H, February 1989. Illinois #6 coal, Burning Star Mine. Based on moisture- and ash-free coal.

tor. Dilution by nitrogen is undesirable, and in the first case relatively pure oxygen must be separated from air. In the second case heat is added to the hot solids or gas in a separate vessel via combustion with air.

The reaction of coal with steam produces, in addition to carbon monoxide and hydrogen, carbon dioxide (CO_2), hydrogen sulfide (H_2S), methane (CH_4), ammonia (NH_3), and particulates. These are currently removed in a low-temperature cleanup train, and when hydrogen is the desired product the gas composition is adjusted by the reaction

$$CO + H_2O = CO_2 + H_2,$$

CO_2 is removed and, where necessary, CO is removed by reaction with hydrogen to form methane.

The energy for water splitting, shifting, and gas cleanup and oxygen generation comes from fossil fuel combustion with additional generation of CO_2. If control of CO_2 emissions becomes necessary, other energy sources could be used. These include biomass, nuclear heat, and solar energy. Biomass could be used to supply heat and hydrogen by modification of technologies developed for coal. The direct use of nuclear heat will require relatively low temperature gasification and would lead to a gasification process specifically designed for integration with a nuclear heat source (see Appendix K).

In the period between the oil price increases of the 1970s and the petroleum price collapse of 1983, there was a wide variety of R&D on hydrogen production processes with the anticipation of extensive use of these processes starting in 1990. Uncertainty about future oil prices has resulted in abandonment of many of these programs. While no commercial coal liquefaction plants were built, a few other applications of gasification are providing commercial experience on coal gasification in this country. Of particular interest is its use in conjunction with electric power generation.

The use of coal gasification for electric power generation is based on the ease of removal of pollutants from the gas and on its potential for use in more efficient combined-cycle systems where the gas is first burned at pressure in a gas turbine followed by steam generation by the hot gas turbine exhaust. The pioneer Cool Water demonstration program in Daggett, California, which makes use of the Texaco entrained flow gasifier, is considered quite successful, and both Texaco and Shell are actively developing and marketing this type of gasifier where finely ground coal is reacted with oxygen at pressures of 150 to 600 psi in a high-velocity reactor with cocurrent flow of coal and gas. Temperatures approach 1650°C (3000°F). Lurgi countercurrent moving bed gasifiers are in commercial use and have demonstrated good performance at the Great Plains coal gasification plant in Beulah, North Dakota.

Assuming there is a continuing market for specific gasifiers, it is expected that the continuing industrial R&D programs will leave little incentive for DOE to engage in research aimed at evolutionary improvements in these specific systems. While these systems have the advantages of being available and generally applicable when hydrogen or syngas is needed, they suffer from problems of durability, small scale, and inability to integrate pyrolysis tar and gas recovery with the total liquefaction system.

The Texaco plants at Cool Water, for example, have a capacity of 1000 tons/day of coal. A design based on Wilsonville technology proposed use of larger-scale gasifiers (6760 tons/day [wet]). A 100,000-bbl/day plant would require 12 of these gasifiers, 9 on-line and 3 standby. While multiple units are needed to allow for periodic repair of ceramic liners, it appears that much larger units may offer opportunity for cost savings for a scale of operation consistent with the scale needed for producing transportation fuels. While these commercial systems could probably be increased in size, other systems may be more amenable to major scale-up and integration with the coal liquefaction process.

The separation of oxygen from air is a major expense in current commercial systems, and the alternative of supplying heat from separately heated solids or gas can be competitive. The sources of heat in this case would be fuel combustion by air or, perhaps in the long term, nuclear heat. These systems, in general, would operate at lower temperatures (below ash fusion) than the oxygen-entrained flow systems. This reduces the problem of ceramic life but introduces the problem of disposing of unfused ash, which would be more easily leached than fused ash. Production of by-product methane increases as the temperature is reduced, and in the presence of catalysts large amounts of methane are generated in situ, as in the Exxon catalytic gasification process. Methane formation is a secondary reaction that can be limited by short gas contact time or by the presence of a CO_2 acceptor such as calcium oxide. Use of fluidized bed technology appears to offer advantages for large-scale solids-handling processes and potential for use in multiple beds and stages.

Many variations of multiple bed systems for coal gasification have been investigated (Kuo, 1984). Figure 4-2 is a generic scheme, which provides for coproduction of hydrocarbons from pyrolysis and can supply heat for the endothermic char-steam reactions. The gasification reactor can produce large amounts of methane, if conditions are chosen to encourage catalytic gasification (as in the Exxon catalytic gasification process). Hydrogen of purity sufficient for coal hydroliquefaction could be produced, under different conditions, where circulated calcium oxide acts as a CO_2 acceptor.

An additional set of alternative processes involves the use of hot iron or iron oxide to produce hydrogen from steam. Iron oxide is reduced by producer gas from reaction of coal with an air-steam mixture. The reduced iron oxide (FeO) reacts in a separate vessel with steam to form hydrogen

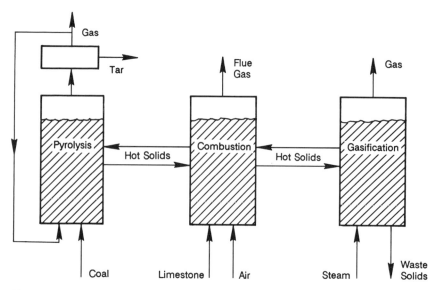

FIGURE 4-2 Schematic of a coal gasification process following pyrolysis and combustion to obtain higher thermal efficiencies.

and Fe_3O_4. A variation involving use of molten iron is being studied by the Japanese.

Underground gasification of coal has been extensively studied in Russia and in the United States. Early U.S. field tests used air injection to produce a relatively low BTU product gas. Injection of oxygen and steam as gasification agents was shown in later tests to substantially improve product heating values. A successful pilot scale test was carried out in a steeply dipping seam in 1981 by a joint project between DOE and Gulf Research and Development Company. This test (Rawlins II) produced good-quality synthesis gas for 65 days at a rate of 130 tons/day. The work was terminated for lack of funds (Singleton, 1982). Estimated costs, using the test results, indicated a substantial reduction in synthesis gas cost (Schulman and Biasca, 1989) (see Appendix D, Table D-3). Because the steeply dipping coal seams qualifying for this technique are considered "unmineable," a very low coal value was used.

The most recent underground gasification field test, the Rocky Mountain I test, was conducted in 1987 to 1988 by a consortium of the DOE and several private concerns headed by the Gas Research Institute. This test demonstrated successful long-term gasification of flat seams via the use of the Controlled Retracting Injection Point (CRIP) technology developed by the Lawrence Livermore National Laboratory (Cena et al., 1988). The Rocky Mountain I CRIP module gasified over 11,000 tons of coal in 93 days to produce a product gas with an average heating value of 287 BTU/standard cubic foot. The gasifier operated at efficiencies in terms of energy recovery per unit of oxygen/steam injected that were comparable with surface gasifiers. With the potential of significant reductions in capital and mining costs compared with surface gasifiers, underground gasification is a promising technology for liquids production from coal, but several questions remain, including

- scaling-up in length of time and rate of gas production,
- effects on aquifer quality for a large operation,
- ability to predict and control performance for less ideal coal seams, and
- evaluation of the amount of coal meeting the requirements of this technique.

A joint industry-DOE commercial demonstration designed to answer the above questions is recommended pending the environmental results from Rocky Mountain. The syngas produced from this demonstration could be used to produce ammonia or methanol.

The DOE Program

Table 4-2 summarizes the 1988 and 1989 appropriations for work relevant to coal gasification.

Advanced Research. The programs on gasification and pyrolysis chemistry, ash agglomeration, and gas separation are important areas. New programs with particular relevance to integrated direct coal liquefaction systems should be established.

Systems for Synthesis Gas Production. The systems development and modeling effort is essential to identification of the optimized gasification approach and the corresponding research thrusts. The scope should probably be extended to include the total system of liquefaction, gasification, and coproducts.

Systems for Coproduct Production. This large program appears to be oriented toward stand-alone pyrolysis liquid-char producing systems. No reference is made to processing pyrolysis liquids in the coal liquefaction reactor in combined pyrolysis-gasification-direct liquefaction systems. As described in the 1990 proposed program, the program is aimed at directly

TABLE 4-2 The 1988 and 1989 Appropriations for Work Relevant to Coal Gasification

	Appropriations ($1000s)		Relevance to Liquid Fuel Production from Coal
	1988	1989	
Advanced research	2,698	2,679	High
Systems for synthesis gas production	1,958	2,679	High
Systems for coproducts production	5,292	9,084	Medium
Total	9,948	14,442	
Underground coal gasification	2,777	1,371	High
Systems for power production	11,176	5,926	Low
Systems for industrial fuel gas	1,369	848	Medium
Great plains coal gasification (methane production)	500	517	Low
Total	15,822	8,662	

marketable nontransportation products. The relevance to major transportation fuels should be improved.

Underground Coal Gasification. As discussed previously, this process has potential for reduced cost from the use of low-value coal.

Production from Biomass. The 1990 DOE Solar Energy Program includes $1.1 million to begin small-scale biomass gasification research. While aimed at methanol production, the same processes could be used to supply hydrogen for manufacture of transportation fuels from fossil resources. Synthesis gas or hydrogen produced from biomass has potentially no net CO_2 effect on the atmosphere if this synthesis gas is used to manufacture methanol or F-T gasoline. If such production of methanol were to occur in the economy, the amount of coal-based gasoline would be less, resulting in less CO_2 output. The alternative of using biomass-produced hydrogen for coal liquefaction, however, results in a larger reduction in coal-produced CO_2 since more liquid fuel is produced per unit of hydrogen (synthesis gas) by the coal liquefaction route than by manufacture of methanol from synthesis gas.

Systems for Power Production. Much of the work on systems for power production is quite specific to utility problems. Hot gas cleanup (H_2S removal) is less important for coal liquefaction since the process is sulfur tolerant and since H_2S must be removed from the spent hydrogen stream. Hot CO_2 removal with concurrent shifting might be of greater interest.

Systems for Industrial Fuel Gas Production. The program on systems for industrial fuel gas production aims for lower-cost oxygen production and operation of the Morgantown Energy Technology Center (METC) fluid bed gasifier. This work can potentially make contributions to gasification for hydrogen and synthesis gas production and should be managed with this in mind.

Conclusions and Recommendations

Manufacturing of hydrogen and synthesis gas is a major economic and energy cost and source of CO_2 in the production of liquid transportation fuels from natural gas and coal, shale, heavy oils, and biomass. Processes for syngas manufacture *from natural gas* are widely used and of continuing R&D interest to industry. The participation of DOE should be limited to exploratory and fundamental studies that are relevant to manufacture from both natural gas and coal.

When petroleum prices increase to the point where coal and shale liquefaction are competitive, it is expected that increases in the price of natural gas will cause a shift to coal gasification. Other sources of hydrogen (biomass, electrolysis, photolysis, thermal water splitting) are expected to be more costly. High priority should, therefore, be placed on reducing the cost of hydrogen and synthesis gas manufacture from coal, reducing CO_2 generation by using nonfossil sources of energy for process heat and hydrogen production, and improving energy efficiency.

While general-purpose coal gasification processes are now in limited commercial use, it is believed that important opportunities exist for development of gasification processes specifically chosen for integration with direct coal liquefaction to reduce both cost and CO_2 production. Such an optimized process might incorporate features such as coproduction of pyrolysis liquids and low-cost methane, larger-scale equipment, and use of air combustion, biomass combustion, or possibly nuclear heat in the long run. Demonstration and broader evaluation of underground coal gasification are also recommended.

The 1989 DOE program is, in general, well chosen; however, it is recommended that increased emphasis be placed on identifying opportunities for reducing transportation fuel cost and CO_2 output by integration of direct coal liquefaction and biomass processes in an era when natural gas prices are high. When such opportunities are identified, the program should be expanded with the goal of working toward a demonstration.

HEAVY OIL CONVERSION

The carbon rejection and hydrogen addition processes for heavy oils are difficult primarily because of the high concentrations of contaminants like sulfur, nitrogen, metals (mostly nickel and vanadium), and coke-forming molecules (known as "carbon residue") in the heavy oils. Metals poison catalysts and reduce upgrading efficiency, and sulfur and nitrogen must be removed in a cost-effective and environmentally acceptable manner. Many of these processes for converting heavy oil or crude oil vacuum residuum (vacuum resid) are commercial and standard in the petroleum industry. The petroleum sector is profiting from early investment and licensing of these procedures. A brief description follows (for more details, see Appendix E).

Commercial Processes

There are a number of commercial carbon rejection processes that upgrade heavy oil to liquids, coke, and gas, the liquids generally of a poor quality. The liquids must usually be hydrotreated before being used as reformer or fluid catalytic-cracking (FCC) feeds to make transportation fuels. These processes include the following:

1. Delayed coking—heavy oil or vacuum resid is thermally cracked in a vessel yielding liquids, gas, and high-sulfur coke.
2. Fluid coking—heavy oil is thermally cracked in a reactor containing a bed of fluidized coke particles. Sulfur oxides (SO_x) need to be controlled.
3. Flexicoking—an extension of fluid coking, in which most of the coke is gasified to low-Btu gas and the need for a coke market is eliminated. Sulfur is removed as hydrogen sulfide.
4. Resid FCC and heavy oil cracking—resid is fed to a fluidized bed with a cracking catalyst, yielding gasoline-range boiling materials with carbon residue deposited on the catalyst. Since heavy metals may poison the catalysts, upstream processing of the resid is usually required.

Commercial hydrogen addition processes include catalytic or thermal hydrocracking, or the donor solvent type. They include (1) fixed-bed residuum or vacuum residuum desulfurization (RDS/VRDS), developed 20 years ago, is increasingly used as heavy oils become heavier. In this process atmospheric or vacuum resid oil contacts catalyst and hydrogen, removes most of the metals and sulfur, and creates an acceptable feedstock for further upgrading in an FCC. (2) Bunker flow or hycon process is similar to RDS/VRDS, except that the catalyst can continuously be added and removed. (3) Ebullating bed processes, known as LC-fining or H-oil, involve hydrocracking and remove metals and sulfur of any heavy oil. The distillate products are of low quality and require further hydrotreating and upgrading.

Processes with Limited Commercial Application

Asphalt residue treatment (ART) is a carbon rejection process with a reactor similar to an FCC. The feed contacts a high-temperature solid and is volatized, and the coke is burned off the solid in a regenerator to produce the required heat. The liquid product yield is high but requires further upgrading.

Hydrogen addition processes include many slurry hydrocracking processes, a variation of thermal high-pressure hydrocracking. A dilute slurry is added to a cracking reactor to suppress coke formation and attract metal contaminants. Conversions of vacuum resid are high, but the products are high in sulfur and nitrogen, requiring further hydrogenation.

Fuel Properties

The product qualities resulting from the various heavy oil upgrading technologies are quite variable and are strongly dependent on feed type, process type, and processing conditions. However, producing fuels of acceptable properties is possible (in all cases) with existing petroleum processing technology, although the economics vary with a given refinery's situation, the feedstock, and product prices.

Environmental Considerations

Air emissions can be controlled as necessary for all of the technologies (see Appendix E). All hydrogen addition processes reduce nitrogen and sulfur contaminants to ammonia and sulfur (via hydrogen sulfide), which are low-value by-products sold for fertilizer and sulfuric acid manufacture.

Solid wastes consist mainly of spent catalyst. Spent resid hydroprocessing catalysts are generally quite high in nickel, vanadium, cobalt, and molybdenum and constitute decent "ores." Metals are frequently extracted and recycled. Spent FCC catalysts are suitable for landfill in most locations but in the future may require disposal as hazardous waste. The Petroleum Environmental Research Forum (PERF) is currently researching the incorporation of spent FCC catalyst in cement, concrete, and asphalt. A few refiners already dispose of their spent FCC catalyst by that method.

The wastewater treatment implications of the newer generation of heavy oil conversion processes are much more complex, and generalizations are difficult.

Opportunities for Cost Reduction

For well-demonstrated processes such as delayed coking or fluid coking, the potential for significant technology improvements and, consequently, for cost reductions is somewhat limited. For the other processing routes the potential for improved catalysts, equipment design, and processing conditions is good, and there is continuing progress in decreasing the capital investment and operating costs.

In 1988 over 1.0 million bbl/day of residual fuel oil was consumed in U.S. electrical, utility, industrial, and commercial boilers. Many refineries are already equipped to convert heavy oils and residua to transportation fuels. However, most refineries use coking processes rather than the more expensive hydroprocessing technologies. Technology advances will allow more extensive use of hydroprocessing, which produces more transportation fuel than coking, especially through conversion of heavy feedstocks that are difficult to process due to high levels of metals, coke-forming molecules, and other aspects.

DOE Research Program Recommendations

The technical literature indicates that the Japanese government is deeply committed to funding heavy oil R&D through such organizations as Research Association for Residual Oil Processing (RAROP) and the Research Association for Petroleum Alternative Development (RAPAD). It is important for DOE to fund appropriate heavy oil research if the United States is to maintain a leadership position in petroleum processing technology. Funding the work in academic laboratories also serves the need to educate more U.S. scientists and engineers to ensure U.S. industrial competitiveness in the future.

The committee recommends funding research on the fundamental chemistry and kinetics of chemical reactions occurring in the existing heavy oil upgrading processes. For example, understanding the mechanisms of coke formation and how reaction pathways might be altered to reduce coking tendency is important. Such fundamental knowledge could be useful in improving many existing processes, both carbon-rejection and hydrogen-addition approaches.

There is very little fundamental molecular information available about the structures of metals, sulfur, and nitrogen-binding sites and coke precursor species in heavy oil feeds and upgraded products. Such information would be very useful in developing new processes and improving existing ones. Such projects are appropriate for DOE funding in academic laboratories where expertise in sophisticated analytical instrumentation exists.

For all hydrogen addition processes the cost of hydrogen is quite a significant fraction of the total upgrading cost (Fant, 1973). The total cost of hydrogen to convert residuum to transportation fuels is in the range of $2 to $3/barrel. Although hydrogen manufacture has been researched for many years, a breakthrough in this area would greatly reduce the costs associated with many heavy oil upgrading processes. RAROP is also building a pilot plant to study inexpensive hydrogen production (*Japan Chemical Week,* 1986).

Further research in the environmental area may be appropriate in the areas of wastewater from newer heavy oil conversion processes. Also, improvements might be made in the extraction of metals from spent hydroprocessing catalysts.

The committee does *not* recommend DOE funding of research on process or catalyst development in the area of heavy oil conversion. There is already an extensive commitment to R&D in this area in the private sector, and much duplication would likely result.

TAR SANDS RECOVERY AND PROCESSING

Tar sands are defined as "any unconsolidated rock containing a crude oil which is too viscous at natural reservoir temperatures to be commercially producible by primary recovery techniques"; American Petroleum Institute gravities are generally less than 10° (IOCC, 1982; see also Appendix C).

While U.S. tar sands deposits are not as extensive as Canada or Venezuela, they are significant (see Table C-6). U.S. tar sands are hydrocarbon wetted rather than water wetted like Canadian tar sands. The percent by weight of bitumen of the U.S. resource varies widely, from about 1.5 in Alabama, to 4 to 10 in Utah, to 30 in California—a value that essentially represents heavy oil (Tables 4-3 and 4-4). Porosity, a characteristic of the host rock, ranges from 15 to 40 percent of total rock weight, and oil saturation usually represents about 50 percent of the porosity. Overburden, which varies by site, is an important determinant of the total cost of tar sands recovery, as is the extent of layering of a given deposit, and bitumen properties also vary widely among different deposits. Recovered hydrocarbons would generally need some onsite viscosity reduction to make them pumpable through refining systems. Like heavy oils, high-sulfur bitumen results in high-sulfur coke in conversion processes producing coke requiring SO_x emissions control. Nitrogen content, generally higher for tar sands bitumen than petroleum, can poison refinery catalysts and affect end-product fuel quality. Conversion of tar sands to products requires mining, recovery, and upgrading.

TABLE 4-3 Tar Sands Reservoir Characteristics

	Grade (% oil)	Porosity (%)	Oil Saturation (%)	Overburden (ft)	Thickness (ft)
Utah	4-10	15	23-72	0-500	10-60
Texas	—	30	35-55	1500	15-300
Kentucky	6-10	15	20-70	0-200	10-50
California	(up to 30)	30-40	50-75	0-3200	50-400
Alabama	1.5	6-24	4-56	0-1000	10-300

TABLE 4-4 Bitumen Properties

	API Gravity	Pour Point (°F)	Viscosity (million centipoise)	Sulfur (wt%)	Nitrogen (wt%)
Utah	5-14	95-150	1	0.4-3.8	0.6-1.3
Texas	−2-10	180	20	10.0	0.4
Kentucky	10-12	55	1	1.5	0.4
California	8-17	—	1	3-7	1.2

Mining

The costs of surface mining recovery of tar sands ore depend on the ore body configuration, thickness, and type and the overburden of nonbearing rock. Overburden thicknesses greater than around 300 to 500 ft result in large initial costs for tar sand commercial plants and significantly reduce economic viability. Variation of ore richness within a deposit affects the mining strategy and production costs.

Mining techniques include drag lines and bucket wheel excavators, which are used in Athabasca tar sands in Alberta, Canada, and are quite effective for bulk surface mining of unconsolidated tar sands. Mining of consolidated tar sands, such as those of tar-saturated sandstones, requires drilling, blasting, and removal with power shovels and trucks. Commercial conveyors are available to move either type of tar sand to the processing plant. Newer mining machines, similar to road resurfacing machines for asphalt roads, are mounted on treads and use a rotating drum fitted with tungsten carbide teeth to rip out thin layers of tar sand ore or overburden. Mining machines are particularly suited to multiple seams of consolidated tar sands.

Recovery

Two types of recovery technologies, extraction or retorting, are generally considered for tar sands. In extraction a solvent such as naphtha dissolves bitumen from the host rock at low temperatures. For retorting tar sand is heated to pyrolysis temperature in a retort vessel. For U.S. tar sands extraction is more cost effective than retorting and is the preferred technology. A higher oil-yield retorting process might make retorting more economical for deposits with lower ore richness. Extraction is more cost effective than retorting for tar sands below about 15 percent by weight richness. This richness includes most of the U.S. resources.

Extraction

Depending on the type of tar sand material, either a water-based or hydrocarbon solvent is needed to recover tar sand bitumen from host rock. The Clark hot-water process, using a caustic water solution to emulsify oil from the tar sand particles, is effective for Canadian tar sands because the sand particles are wetted with water and surrounded by bitumen. The process recovered initially about 85 percent of the bitumen. Recent process improvements have increased recovery to between 88 and 92 percent (Coal and Synfuels Technology, 1989a). The unrecovered bitumen is discharged with the tailings, causing a serious environmental concern.

U.S. tar sands, in contrast, are hydrocarbon wetted, and effective bitumen recovery requires a hydrocarbon solvent used in processes specifically tailored for a given process. Figure 4-3 shows a generalized solvent extraction process. Tar sands and solvent come together in an extraction contactor, such as a mix tank, rotating drum, or other mixing device. In this example water is also added to wet the mineral particles after the solvent has had time to wash bitumen from the sand. Separation of the solvent-bitumen phase from the water-sand phase is easier than in the hot water process, since emulsions can generally be avoided with the solvent process. Tailings can be separated in a raked bottom separator by gravity. Using water in this process to displace solvent and bitumen from sand saves energy in recovering solvent from the solid tailings. The bitumen solvent phase is separated, and fine solids are removed in a centrifuge, lamella, or by other means. Solvent recovery is accomplished in evaporators or by distillation to produce a relatively solids-free bitumen. Recycled solvent is recovered for reuse.

Critical features of hydrocarbon solvent extraction include essentially complete recovery of solvent from spent tar sands and good removal of mineral fines from bitumen. Residual solvent concentrations of only 100 ppm represent substantial losses of solvent, causing environmental problems and reduced profitability. The extent of fine solids removal affects the adaptability of existing refinery processes for upgrading bitumen to lighter products.

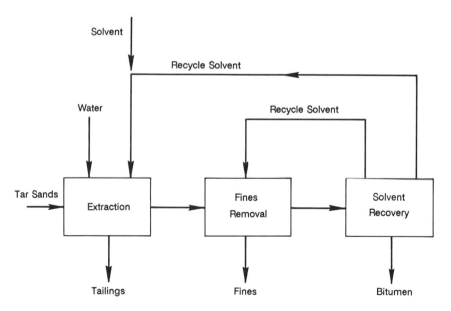

FIGURE 4-3 Hydrocarbon solvent extraction process.

Retorting

Many possible retorts might work for tar sands, but the Taciuk, Allis-Chalmers, Lurgi, and fluid-bed retorts have been the subject of the most recent R&D. The Taciuk retort is under development in Canada by the Alberta Oil Sands Tar Research Authority (AOSTRA) for Canadian tar sands but can be used on U.S. hydrocarbon-wetted tar sands. The Taciuk retort is a rotating drum device, 9 ft in diameter, with efficient, countercurrent heat transfer, which effectively retorts and cracks tar sand to light oils and gases; the feed material is predried by contact with exiting combusted spent sand. A 9-ft-diameter pilot retort has been tested to date. A similar rotating drum retort was under development some years ago by Allis-Chalmers near Milwaukee.

The Lurgi-Ruhrgas retorting process has been commercially used for coal pyrolysis. When using the process with tar sands, the feed is mixed with hot combusted spent sand in a screw conveyor and added to the retorting reactor, which operates at 480° to 510°C (900° to 950°F) (Figure 4-4). The bitumen cracks to gas and light oil, which pass overhead as a vapor, and to coke, which deposits on the spent sand. A cyclone separates sand fines from the product vapors, and the vapors are then condensed to an oil and water stream. The retorted sand flows downward to the bottom of a riser combustor pipe, where air is injected to burn the carbon and unreacted bitumen. The combustor reaches temperatures of 590° to 650°C (1100° to 1200°F). Sulfur and nitrogen emissions in flue gas must be properly controlled. Spent sand can be rejected from the product cyclone or retort as shown, but the preferred configuration is to withdraw and cool spent combusted sand from the cyclone. Fluid bed retorting processes also provide residence time for retorting and burn coke from the spent sand to achieve environmentally acceptable tailings.

Upgrading

Upgrading bitumen to higher H/C ratio products is similar to upgrading heavy oil, either through carbon rejection or hydrogen addition processes. The choice of process depends on oil prices, bitumen reactivity, sulfur, nitrogen content, and the combination of upgraded products desired.

In general, crude products from bitumen upgrading or retorting conversion processes are further upgraded to produce liquid feedstocks for existing refinery units and by-product fuel gas. The naphtha product is hydrotreated to remove nitrogen prior to further refining, to produce high-octane motor gasoline blending stocks. The distillates also need to be hydrotreated to meet diesel, heater oil, or furnace oil specifications. Gas oil is hydrotreated to remove nitrogen so it can be fed to a catalytic cracker ("cat cracker"), which further cracks the gas oil to additional naphtha, distillate,

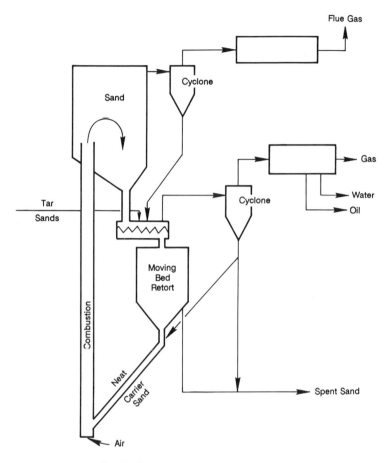

FIGURE 4-4 Lurgi-Ruhr gas process.

and gas. Since sulfur is more easily removed than nitrogen in hydrotreating, the sulfur content of upgraded products from Utah or other low-sulfur tar sands will be very low.

Environmental Considerations

Tar sands processing converts part of the bitumen to coke and fuel gas, which are burned in processing plants and refineries to generate steam and power. Sulfur and nitrogen oxide emissions controls are not expected to be significantly different than those for present refineries. Sulfur oxide control should, in fact, be somewhat easier for very low-sulfur Utah tar sands.

Likewise, environmental controls for water quality related to mining operations do not present any new or special requirements over current practices. On the other hand, new techniques are required to ensure that the spent tailings from retort or extraction operations can be disposed of safely in a landfill. Complete combustion of hydrocarbons on retorted sand will be required. For extraction processes the complete removal of leachable hydrocarbons will be required. Where a solvent is used, complete recovery of the solvent will be critical.

The consumption of fuels made from low-sulfur Utah tar sands should produce lower sulfur oxide emissions, since average refinery products have higher sulfur levels. Nitrogen emission levels are expected to be about the same as for current refinery products.

Economics

The overall economics of a commercial tar sands plant depends on the following, in descending order of importance: (1) resource richness, since richer ores require less ore to be mined and are easier to process through primary recovery; (2) bitumen quality, primarily its nitrogen and sulfur content as well as its type and resulting yield structure from upgrading; (3) reduction of energy process requirements, to avoid purchasing large amounts of electrical power, on the one hand, and expensive capital outlay for power generation equipment, on the other; (4) such resource characteristics as overburden thickness and the ability to mine selectively high-grade portions of deposits; and (5) resource location as it affects markets, transportation, and labor productivity. Using extraction technology and bitumen upgrading through an asphalt residue treatment process, the costs for upgraded tar sand products are about $26/barrel to $35/barrel. The costs for pyrolysis processes are generally higher.

A number of cost reductions for U.S. tar sands processing could improve its overall economics. These opportunities include improved bitumen upgrading; higher bitumen recovery, which would require less mining; improvements in solids removal, which could amount to 15 percent of the total cost of products depending on mineral characteristics; and improved techniques that permit more selective mining of the richest ore, which would reduce both processing and mining costs. The committee judged that these improvements and the use of a rich resource such as Utah tar sands could reduce costs to about $20/barrel of crude equivalent (based on the 10 percent discount rate cited in Chapter 3).

Recommended Areas of Research for DOE

Based on the above considerations, a broad R&D program for exploiting domestic tar sands should include the following areas:

- characteristics of nitrogen compounds, to improve the upgrading of low-API-gravity, high-nitrogen bitumen;
- solid-liquid interactions, to improve removal of fine solids;
- dissolution mass transfer effects to optimize the extraction and separation operations;
- improved extraction techniques, to improve bitumen recovery;
- upgrading of feeds containing solids (need pilot plant demonstrations) due to solids handling requirements for disposal;
- environmental studies, especially for solid waste disposal; and
- improved techniques for efficient and selective mining of the richest ores.

The current DOE program for surface extraction of tar sands is focused on the Western Research Institute (WRI) bitumen recycle process and optimization of a few alternative recovery processes. Over the next 5 years these and other available processes along with mining techniques applicable to U.S. tar sands should be evaluated both technically and economically. Engineering firms, petroleum operating companies, and qualified consultants should play a key role in these evaluations, which should be used to determine if any processes are suitable for further development in a field pilot facility having a capacity of 50 to 100 bbl/day. Building on Canadian experience, this size should be suitable for scale-up to a commercial plant. A field pilot operation is justified only if the technology is judged to be sound, all enviromental requirements are projected to be met, and costs are sufficiently low (probably below $25/barrel) to attract industry participation. If a technology worthy of development cannot be identified, serious consideration should be given to terminating the government program. Continuation of the program would be justified only if the prospects for improvement were judged to be outstanding.

OIL SHALE

The Green River formation in Colorado, Utah, and Wyoming is the largest and richest oil shale deposit known. The most valuable part of the resource is in the Piceance Creek basin in western Colorado. The northern and central parts of the basin (600 square miles) contain about 720 billion bbl of shale oil (Lewis, 1980) in thicknesses ranging from 400 to over 2000 ft with a relatively thin overburden. Almost all this thick, rich shale is owned by the government, the main exception being the two federal leases (tracts C-a and C-b) held by Amoco and Occidental Petroleum, respectively. Much of the resource is suitable for very large scale, low-cost open pit mining. This possibility has received little attention, perhaps because most of the land is government owned and is not available for lease or sale. Tract C-a is suitable for such open pit mining.

A smaller area in the southern part of the basin is privately owned, mostly by oil companies. The oil shale of economic interest in this area consists of a thin, high-grade layer of shale, the Mahogany zone, generally 30 to 150 ft thick. This layer of oil shale, averaging 30 gal/ton, contains approximately 50 billion bbl of shale oil (Lewis, 1980). This layer is suitable for room and pillar mining but generally not open pit mining because of the high ratio of overburden to oil shale. In the central part of the basin the Mahogany zone is thicker and richer. The other zones, especially those deeper in the section, also contain economically important quantities of oil.

The Green River formation also underlies large areas of Utah and Wyoming. In these areas the shale that averages more than 20 gal/ton is less than 400 ft thick. High-grade layers in these regions are for the most part too thin or too deeply buried to be of economic interest for the foreseeable future, although in a small area of eastern Utah shale occurs close to the surface. Although the shale in this area is generally thinner and of lower grade than that in the southern part of the Piceance basin, it is otherwise similar and suitable for room and pillar mining.

Deposits of oil shale other than the Green River formation occur in the continental United States and Alaska. The most widespread of these are the oil shales of Devonian to Mississippian age, generally referred to as eastern shales. These eastern shales are thinner and of lower grade than the Green River formation as measured by Fischer assay, a standardized test that measures the amount of liquid oil that can be obtained in ordinary pyrolysis processes. Kerogen in eastern shales has a lower H/C ratio and is more like coal than the Green River shales. For thicknesses of 10 to 30 ft the average grade is usually between 5 and 10 gal/ton by Fischer assay. Increased conversion to liquids by adding hydrogen is possible but requires a more complicated and expensive process (Hu and Rex, 1988). The following sections will concentrate on the shale of the Green River formation, which has by far the most economic potential.

State of Technology Development

Shale Properties and Process

The Green River oil shale is an impure marlstone (a silicate/carbonate rock) consisting mostly of inorganic mineral matter. The solid organic constituent is kerogen, in a typical shale making up about 15 percent of the rock by mass and 30 percent by volume. Due to the nature of the oil shale, any practical physical process to completely separate the kerogen from the mineral matter is unlikely. A less than complete separation might be useful if the beneficiation process is sufficiently inexpensive. Because the raw shale is essentially impermeable to fluids and the kerogen is not soluble, ordinary (subcritical) liquid extraction processes are likely to be unsatisfactory. Pyrolysis or retorting is therefore the process of choice.

In surface retorting shale oil is broken into pieces small enough to permit good heat transfer. (In any shale mining and crushing operation this requirement is easily met, although the size range and content of fine particles present some handling difficulties.) The shale pieces are then heated to about 930°F (500°C) to pyrolyze the kerogen, producing oil, gas, and solid char. Heat for the retorting process is provided most efficiently by burning the solid char formed on the shale mineral during retorting.

Carbonate minerals, dolomite and calcite, are major components of these shales, and dolomite especially may be decomposed at high temperatures. Decomposition of carbonates during processing wastes energy and produces CO_2, which may be desirable if there is a market for CO_2 and undesirable if it proves necessary to limit the production of greenhouse gases. Some reaction of carbonates with SO_2 is desirable to eliminate the emission of sulfur gases. Carbonate decomposition may be avoided at practical retorting temperatures, but some decomposition usually occurs at the higher temperatures used for char combustion. The extent of carbonate decomposition depends on the time and temperature required for combustion.

Shale oil produced by current processes is somewhat unstable, has a high viscosity and a moderate dust content, and usually cannot be put directly into a pipeline. Upgrading has been required to reduce the viscosity and stabilize the oil for transport to refineries. If a hydrotreatment process is used for this upgrading, the sulfur and nitrogen contents are also reduced and a premium refinery feedstock is obtained that contains almost no heavy residual fractions.

The challenge is to devise a simple and efficient oil shale process for mining, retorting, upgrading, and disposal at low cost. Reaching this goal will require R&D and translation to commercial practice.

Mining and Disposal

To a great extent mining technology is transferrable to oil shale. The scale of the operation is large, however, and there are opportunities for improvements. Room and pillar mining on a large scale have been demonstrated in the Anvil Points, Colony, and Union mines in the Piceance basin.

Much of the larger resource of rich shale in the central Piceance basin is ideal for open pit mining, with very low stripping ratios (overburden/ore). This mining would be on a large scale because of the thickness and extent of the resource. Open pits would be 2000 to 3000 ft deep and thousands of feet across. Disposal of waste is included with mining because of similar materials handling and the necessity for an integrated operation.

The ideal process would permit consolidation of waste shale, other solid wastes from the plant, and wastewater. While studies indicate that this approach is feasible, disposal of these materials in an open pit has not been demonstrated. Both spent shale and mined but unprocessed rock (overburden and lean shale) must be deposited in the pit and compacted in a stable configuration. Additional space outside the pit would also be needed, to the extent that the waste solids will not pack as densely as unbroken rock and allow working space during early development of the mine.

For the still large remainder of the resource in the central Piceance basin, where the overburden stripping ratio is too high for open pit mining, another mining method must be found. Room and pillar mining can be used, but as the depth and thickness of the shale increase, the fraction of the resource that is recovered will be less than 50 percent. No underground mining method is now available for oil shale that can recover a larger fraction of the resource. Block caving is a low-cost mining method that has proved successful for massive copper ores. Large-scale block caving, if applicable to oil shale, could result in substantial cost reduction and improved resource recovery. In these deep zones modified in situ processing (like the Occidental Petroleum Process) is a possible way of recovering this resource and may compete favorably with room and pillar mining.

Retorting

Different retorting processes have evolved from different methods for heating solid particles of oil shale. Either hot gas or hot solid material may be used to supply heat to the shale. (Hot liquids are not practical for temperatures near 500°C.)

If hot gas is chosen as the heat transfer medium, the crushed shale is distributed in a bed through which hot gas is pumped. The various types of hot gas retorts have certain characteristics in common. The mass of gas required to process a unit mass of shale ranges from 0.6 to 0.8, owing to the relative heat capacities of the shale and gas. As a result very large volumes of gas are required, which is undesirable because of the expense of compressing gas and the processing time required for gas-solid contact.

Fine particles in these packed bed retorts restrict the flow of gas or increase the pressure drop across the bed. Therefore, shale particles with diameters less than about $1/8$ in. must be discarded or processed in some other way. Large particles, on the other hand, require a relatively long time to heat, adding costs because of the increased residence time of shale in expensive equipment and the loss of oil through coking, which occurs if the heating is very slow. The practical size limit for shale particles in hot gas retorts is probably a diameter of between 2 and 3 in.

The second option for heating the oil shale is to use a hot solid material.

The raw shale is crushed small enough for rapid heat transfer. Either shale itself (after retorting) or another solid may be used for the heat-carrying medium. The solid heat carrier is heated by burning the residual carbon on the spent shale. Appendix F contains a brief description of several hot gas and hot solid retorting processes.

Upgrading

The properties of shale oil vary as a function of the retorting process. Fine mineral matter carried over from the retorting process and the high viscosity and instability of shale oil produced by present retorting processes have necessitated upgrading of the shale oil before transport to a refinery. After fines removal the shale oil is hydrotreated to reduce nitrogen, sulfur, and arsenic content and improve stability; the cetane index of the diesel and heater oil portion is also improved. The hydrotreating step is generally accomplished in fixed catalyst bed processes under high hydrogen pressures, and hydrotreating conditions are slightly more severe than for comparable boiling range petroleum stocks, because of the higher nitrogen content of shale oil. Shale retorting processes produce an oil with almost no heavy residual fraction. With upgrading, shale oil is a light boiling premium product more valuable than most crude oils.

Considerable cost savings would be possible if the oil could be directly transported to a refinery. Hydrotreating costs would be reduced and duplication of operations would be eliminated by using existing facilities. Expanding or modifying refinery components where necessary would be less expensive than building new facilities in the field, and the operating costs at existing refineries are generally lower than the costs at remote locations. New retorting processes may permit this option.

Advanced Retorting Technologies

New aboveground shale technologies are needed to reduce residence time, increase oil yield, improve oil quality, and thereby cut costs. Efforts to develop modified in situ (MIS) retorting aim to cut costs by reducing the mining expense as compared to conventional retorting. If MIS retorting is successfully developed, the shale mined will be processed in aboveground retorts. In general, hot gas retorts are current technology, and hot solid retorts are advanced retorting technologies (Table 4-5).

In addition, hot gas retorts using internal combustion produce a low fuel value gas and have high mineral decomposition compared with high fuel value gas and low mineral decomposition with either hot solid systems or hot gas retorts using external combustion.

For hot gas retorts the shale is crushed to $1^{1}/_{2}$ to 3 in., with shale less

TABLE 4-5 Comparison of Hot Gas Versus Hot Solid Retorting Systems

Hot Gas Retorts	Hot Solid Retorts
Commercial operations	No commercial operations
High throughput of gas	Low throughput of gas
Slow shale heat-up	Rapid shale heat-up
Large shale required	Smaller shale required
Fines discarded	Fines processed
Sulfur released as hydrogen sulfide	Sulfur retained as sulfates
Oil yield below Fischer assay	Oil yield at Fischer assay

than $1/8$ in. rejected. Hot-solid retorts require a top size of $1/4$ in., with all shale including fines used in some process schemes.

Hot Recycled Solid (HRS) Processes. Virtually all proposed advanced oil shale retorting systems use hot recycled shale as the heat carrier, providing rapid mixing with raw shale, rapid heating, and a subsequent soak time of 1 to 2 min for pyrolysis to occur. This process greatly increases throughput and reduces costs, compared to hot gas processes. An HRS process uses the following major components: a raw and recycled solids mixer, a soak tank pyrolyzer, a pneumatic transport, and a combustor. Many mixer types have been proposed, including fluid beds, gravity fall units, and various mechanical mixers. A requirement for both mechanical and gravity fall units is very rapid mixing (in less than 10 to 20 s) in a compact space (high product space velocity), to avoid excessive oil loss by cracking. In fluid beds longer mixing times are acceptable because oil vapor is quickly swept from the bed. However, more difficulty in condensation of the oil vapor and loss of fines by elutriation are disadvantages of this method.

The soak tank pyrolyzer can be either a moving packed bed or a fluid bed. In some designs the mixer and pyrolyzer are combined into a single fluid bed unit. Disadvantages of a fluid bed pyrolyzer are the expense of gas compression and cooling and the problem of nonuniform and nonoptimum solid residence time. Experiments have shown a moving packed bed pyrolyzer to produce a lighter, less viscous oil that is relatively stable after cooling (Cena and Mallon, 1986). This method may allow direct transport to a refinery without field upgrading, a major cost reduction. In addition, moving packed beds have the advantage of processing essentially all the shale because fines remain in the bed and are retorted.

An air pneumatic lift pipe transports the shale upward to the combustor. A lift pipe high enough to complete combustion is difficult to operate with expected variations in shale grade and reactivity so it may be desirable to combine a fluid bed combustor with a lift pipe. Another system uses a delayed-fall combustor downstream of the lift, which retards the downward fall of solids while air is blown upward or crosswise, providing in a short height the required residence time (10 s) for combustion.

Information from process development or pilot-scale units is not available. Chevron briefly tested a fluid bed HRS process at a scale of 150 tons/day at its Salt Lake City refinery in the mid 1980s. Exxon also fielded a pilot test of its own design at its Baytown refinery during this time. Results of both tests were rumored to have been successful but are not publicly available. The largest tests of the process in the public domain were conducted from 1984 to 1988 by DOE at the Lawrence Livermore National Laboratory (LLNL) using a 1-ton/day solid recycle retort operated from 1984 to 1988. These experiments allowed study of chemical reactions important in the process and tested both a fluid bed and gravity bed pyrolyzer and a lift pipe and delayed fall combustor system. Substantial differences in oil properties were observed between these two pyrolyzers, with major improvements in oil properties observed in the gravity bed pyrolyzer. Reactions occurring between the gas and solids (some catalytic) convert gas to liquid, crack the liquid, and reduce the viscosity of the shale oil (without loss in liquid yield). These reactions are not yet well understood, but the possibility exists to produce shale oil that can be transported to a refinery without upgrading. The LLNL retort is being modified to process 4 ton/day, which will allow tests using the full particle size range of commercial plants and permit study of some important solids handling questions.

Environmental Considerations

Advanced retorting techniques, using hot recycled solids, offer distinct advantages for controlling gaseous emissions. Pyrite forms hydrogen sulfide in the pyrolyzer in both hot gas and hot solid retorts. However, Fe_2O_3 in recycled shale scavenges hydrogen sulfide (H_2S) in hot solid systems, reducing H_2S concentrations to under 1000 ppm. In the combustor FeS burns to form SO_2. Here carbonate minerals scavenge SO_2 at combustor temperatures below 700°C, and at these low temperatures nonfuel NO_x emissions may be avoided. NO_x emissions from the fuel are low. Further reduction may be required and may be possible with additional study of the nitrogen chemistry. More complete combustion may be necessary to reduce CO emissions. In any case the quantity of gas requiring cleanup is generally much less in hot solid than in hot gas retorts.

Use of any fossil fuel contributes to release of CO_2 (see Chapter 5). The amount of CO_2 released per megajoule of useful power is process dependent and varies widely. Western shales contain an average of 40 wt% carbonate minerals, which will decompose if held too long at elevated temperatures. Hot solid processes can be designed to minimize the time that carbonate minerals see elevated temperatures. Mathematical model calculations and small pilot experience indicate that carbonate decomposition can be held below 10 percent in a typical HRS process, thus reducing CO_2 formation.

Waste shale as it comes from the retort must be cooled, transported, and disposed of in the mine or elsewhere. In concept the waste will be cooled, moistened, and compacted into a strong impermeable mass that will stay in place and not contaminate groundwater. Waste shale with part of the carbonate decomposed is a natural cement and will set up after disposal, forming a low-permeability material with adequate strength.

Experience in both wastewater treatment and shale disposal awaits development of process schemes and demonstration tests. Unocal has demonstrated successful methods and experience in these areas for hot gas processes. Similar data are needed for hot solid processes on a comparable scale.

Less direct environmental impacts include the growth and development of the area, with more people, more activity, more roads, and more pollution. The construction of mines and processing facilities will not disrupt other human works, because the area is largely uninhabited; but it will affect natural scenery and wildlife.

Development of water resources and use of water are always important issues in the western states. Shale oil production of 1 to 2 million bbl/day is possible without importing water into the basin (Brown and Stewart, 1978; Sparks, 1974). Larger-scale production would probably require water to be imported but at a cost estimated to be a small part of the total cost of shale oil production. Requirements for prevention of significant air quality deterioration in some areas will also be important. The present prevention of significant deterioration requirements, both federal and state, particularly in Colorado, could limit the production of oil from shale.

Potential for Cost Reductions in Oil Shale Processes

Numerous studies have been conducted to assess the profitability of producing oil from shale. Large variations in the results depend on assumptions about end use, scale, resource type, equipment, contingencies, and risk factors.

The endogenous price calculation in Chapter 3 for oil shale indicates a price of about $43/barrel of crude oil equivalent. Potential cost reductions in the major categories involved in oil shale production and conversion are given in Table 4-6, not including those accruing from making it possible to transport shale oil to a refinery by pipeline without upgrading. Economic studies of both surface retorting and of combining MIS with HRS above

TABLE 4-6 Potential Cost Reductions for Oil Shale Conversion

	Improvements in Current Technology		Developing Advanced Technology	
	Total Capital Cost Reduction Range (%)	Total Product Cost Reduction Range (%)	Total Capital Cost Reduction Range (%)	Total Product Cost Reduction Range (%)
Mining	2-6	3-5	2-11	3-9
Retorting	3-6	3-4	8-20	5-15
Upgrading	2-4	1-3	8-13	6-10
Environment	1-2	1	2-3	2-3
Total Cost-Savings Potential	8-18	8-13	20-47	16-37

surface retorting systems indicate costs in the high $20s to high $30s/barrel of oil produced (U.S. DOE, 1989c; Piper and Ivo, 1986). These results, coupled with the possible cost reductions in Table 4-6, indicate that a cost target of $30/barrel or less is possible with development of advanced technology.

Recommendations for DOE Research Program for Oil Shale Development

The annual DOE program in oil shale is about $10.53 million for fiscal year 1989 with both eastern and western oil shale included. Of the total, $7.46 million is for technology base studies with about $2.6 million of this for eastern oil shale. Environmental mitigation studies constitute about $3.07 million of the budget. Development of oil shale requires substantial lead time and steady progress toward demonstration of promising technologies. The private sector has greatly reduced efforts to develop oil shale technology in favor of developing and producing petroleum and protecting short-term profitability. The major oil companies in particular perceive better near-term opportunities in foreign exploration and production. Government ownership of the thickest and richest oil shale resource is also a factor in discouraging private investment. Few if any companies have sufficient resources to justify more than one or two plants, which probably could not realize the economic benefits of improved technology. At the same time it is not feasible for government to sell or lease substantial parts of its holdings until an industry is started and a value established. Government involvement is therefore essential to advance oil shale technologies to the next stage of development.

Technology development in this field will benefit most from the appropriate mix of fundamental and applied research, process development and scale-up on a step-by-step basis, and industrial experience in operating plants that use full-scale components of all process steps. Much of the knowledge gained in the past has been lost as efforts were halted, yet much expertise remains to provide a base for a renewed effort.

Research Areas

Materials Handling and Solids Flow. The major obstacle to efficient scale-up in the oil shale industry is the handling of bulk solids. Throughout industry, processes that handle solids are prone to long start-up times and operating problems. Fundamental studies of granular flow and materials handling are needed to reduce the costs of oil shale processing.

Reactions and Kinetics of Kerogen Retorting. More work is needed to understand the fundamental chemistry of oil shale retorting, especially specific reactions and kinetics, to solve problems and reduce costs. Additional work is needed in pyrolysis and combustion chemistry to better design and optimize the retorting process and equipment.

Vapor and mineral interactions in the retort affect oil yield and such oil properties as viscosity and stability. Field upgrading might be eliminated if the coking and cracking reactions between oil vapor and solid surfaces are better understood and shale oil with low viscosity and stable enough for pipeline transport can be produced.

Nitrogen and sulfur chemistry are important in reducing NO_x and SO_x emissions and also in cooling of waste shale with water and eliminating hydrogen sulfide release. Decomposition of some carbonate minerals and cement-like reactions occurring during and after disposal are important to efficient spent shale disposal and prevention of leaching of waste and contamination of groundwater and surface streams.

Process Application Chemistry. Process chemistry of waste cooling, waste solid consolidation, and leaching is important to reduce process costs and prevent environmental damage. Studies are needed of reactions between water and minerals during cooling, disposal of solids and wastewater, and exposure of waste solids to groundwater.

Removal of fine mineral matter from product oil is the first step in upgrading. Upgrading of shale oil is usually a process applied to oil after retorting. It may also occur as a result of the primary conversion process in the retort. Research on advanced upgrading processes should be done.

Improved methods for removal of solids from vapor and liquid oil are also needed. Advanced processes for downstream upgrading include better hydrotreating catalysts, and catalysts, absorbents, and solvent extraction or other means for nitrogen removal. Methods are also needed to stabilize shale oil and reduce its viscosity to allow direct shipment to refineries.

Modified In Situ Processes. Research applicable to MIS processes includes gas cleanup, CO_2 removal from the gas stream (if reducing CO_2 emissions becomes important), and understanding and preventing the contamination of groundwater from mining or leaching of spent shale in underground retorts.

Mining Research. Room and pillar mining and open pit mining are adequate for a large part of the oil shale resource during early development. In the longer term a method of underground mining, perhaps some kind of block caving, should be developed to recover a large fraction of the oil shale in the north-central part of the Piceance basin that is too deep for open pit mining. Research on groundwater contamination by mining also is important.

While laboratory experiments and calculations (including mathematical models) are essential, in many cases the important process phenomena can be detected or studied only in the field on a larger scale. Research must therefore be conducted in the field in pilot plants as well. Advanced retorting technologies (hot recycle solid processes) are ready for small-scale field pilot plants. Solids handling, gas isolation methods, continuous operation, proof of concept, materials and durability, scale-up design, and cooling and disposal of waste shale can be studied at this scale.

Timetable for Development of Oil Shale Technology

Oil shale will be a potential source of large amounts of liquid fuel for a very long time. Improvements in technology and industrial experience can make the resource competitive. A long lead time is required to develop technology and to obtain substantial production capacity. Development of advanced technology for a demonstration plant could be accomplished in a decade. From the decision to proceed, 2 years would be needed to design and construct a field pilot plant or process development unit (perhaps 100 bbl/day) for surface retorting. Two more years would be required to operate the plant and plan a full-size pilot plant (1000 bbl/day). An additional 5 years would be required for construction and successful operation of the full-size pilot plant and design of a demonstration plant. With this timetable the United States could have the option by the year 2000 to proceed with construction of a demonstration commercial module (5000 to 10,000 bbl/day) that could take 3 to 5 years from construction to operation. This will demonstrate the technology and allow some cost reduction through operating experience. Further cost reduction can then be expected with construction and operation of a first commercial plant, which would take another 3 to 5 years. Thus, the United States could be in a position to begin to produce shale oil with low-cost advanced technology by 2010. Demonstration of combined in situ (MIS) and surface retorting technology takes the same amount of time, with the critical path being surface retorting. Based on the scenarios presented in Chapter 1, the cost of crude oil could exceed $30/barrel before an advanced oil shale technology is demonstrated.

Because of the long lead time required, government ownership of the land, and industry attraction to short-term opportunities for petroleum exploration and development overseas, industry is not willing to aggressively develop this technology at present. It seems appropriate for the government to take the lead in its early development. However, it is important for industry participation to grow at each scale-up step. Successful demonstration will require the full participation of industry on a cost-shared basis. The decision to proceed with each step will also depend on the success of the technology in cutting costs and a reevaluation of future petroleum prices. The option to proceed with a demonstration plant by the year 2000 requires an accelerated R&D program in comparison to the current one. The committee also judges that, because of the nature of the resource, development of eastern shale should not be conducted at this time.

SYNGAS-BASED FUELS

Syngas can be converted to liquid fuels in two ways: (1) conversion to methanol, which can be followed by conversion to gasoline and distillate, and (2) via the Fisher-Tropsch (F-T) process to hydrocarbons.

Methanol from Syngas

Methanol is made by the catalytic conversion of syngas at about 250°C and 60 to 100 atm. The current commercial processes use a fixed bed catalytic reactor in a gas recycle loop. There are a wide range of mechanical designs used to control the heat released from the reaction and the temperature profile so as to increase per pass conversion. Lurgi and ICI, Inc. (Imperial Chemical Industries) technology dominate. Other designs are offered by Mitsubishi, Linde, and Toyo corporations. New developments in methanol technology include the following:

• *Fluidized bed methanol synthesis being developed by Mitsubishi Gas Chemical.* In this design a fine catalyst is fluidized by the syngas. Better contact between syngas and catalyst gives a higher methanol concentration

exiting the reactor, which reduces the quantity of recycle gas, the recycle compressor size, and the heat exchanger area in the synthesis loop.
- *Liquid-phase slurry reactor for methanol synthesis.* This DOE-supported effort is being developed at LaPorte, Texas, by Air Products and Chemicals, with technical assistance by Chem Systems, Inc. It is estimated that the Mitsubishi technology will reduce capital investment and natural gas usage by 7 and 6 percent, respectively.

Methanol-Derived Fuels

Methanol to Gasoline (MTG)

Methanol can be converted to gasoline using the MTG process developed by the Mobil Research and Development Corporation. The Mobil MTG process is operating commercially in New Zealand, using the ZSM-5 catalyst in fixed bed reactors, to produce 14,500 bbl/day of gasoline. A fluidized bed MTG reactor has been demonstrated by Mobil in a 100-bbl/day semiwork plant in conjuction with West Germany. This program was jointly conducted by Mobil, URBK (Union Rheinische Braunkohlen Kraftstoff AG), and Uhde Gmbh, with additional funding from DOE and BMFT (the Bunden Minister fuer Forschung und Technologie). Also, a direct heat-exchanged MTG reactor concept has been developed by Lurgi.

Haldor Topsoe and Mitsubishi have developed processes that are combinations of methanol (or other oxygenate) synthesis and MTG synthesis. The Advanced Mitsubishi Synthesis Gas to Gasoline process has been demonstrated at the pilot plant scale (1 bbl/day). The Haldor Topsoe process (called TIGAS-Topsoe Integrated Gasoline Synthesis), which is similar to Mitsubishi's process, is based on a mixture of methanol and dimethyl ether as an intermediate. The TIGAS process has been demonstrated at the semicommercial scale in Houston.

Methanol Conversion to Olefins and Diesel

During the MTG development work at Mobil, it was discovered that the yield could be shifted to light olefins by varying the process conditions. The 100-bbl/day fluidized bed semiwork plant in West Germany was also used to demonstrate the methanol-to-olefins (MTO) mode of operation. High-quality gasoline is also produced. Lurgi has also reported MTO pilot plant results using commercial catalysts.

Using olefins from the MTO process (and the F-T process described below) diesel and gasoline can be produced. Catalytic polymerization is a standard refinery process using acid catalysts and is currently being used at SASOL to convert C_3-C_4 olefins to gasoline and diesel. Recently, Mobil has developed an MOGD (Mobil olefin to gasoline and diesel) process using a Mobil commercial zeolite catalyst. The primary products are methyl-branched isoolefins, which have good octane ratings in the gasoline range. The diesel-range olefins can be hydrogenated to give isoparaffins, which have excellent diesel properties. A commercial-scale test of the MOGD process was successfully conducted at a Mobil refinery in late 1981 using feedstock from an FCC unit.

Methanol for Electricity Generation

Coal gasification is currently being developed to generate syngas for power generation. These facilities are referred to as integrated gasification combined-cycle (IGCC) power plants. Although syngas is used directly as a fuel, methanol can also be produced as a coproduct. The production of a methanol coproduct in an IGCC system could have the following advantages: (1) making available a clean liquid fuel for use in peaking service or for sale; (2) increase flexibility of service; and (3) level the IGCC plant at constant operation (95 percent on-stream factor) using methanol for cycling duty.

F-T Synthesis and Product Upgrading

The F-T process is a nonselective polymerization process that produces a range of products, including light hydrocarbon gases, paraffinic waxes, and oxygenates. Further processing of these products is necessary to upgrade the waxy diesel fraction, the low-octane-number gasoline fraction, and the large amount of oxygenates in the product water.

Commercial F-T Processes

Hydrocol Process (Hydrocarbon Research, Inc.). The only U.S. commercial natural gas-to-liquids F-T facility was operated at Brownsville, Texas, from 1950 to 1957. The plant produced an olefinic gasoline with a small quantity of diesel fuel. The plant was shut down in 1957 because of the abundance of low-price crude.

SASOL. The Lurgi/SASOL Arge fixed bed process has been operated commercially at Sasol I in South Africa for about 30 years and, using an iron catalyst, produces predominantly a waxy product. The Sasol Synthol circulating fluid bed process also uses an iron catalyst but operates at a higher temperature. Primarily light olefins and olefinic naphtha are produced. This process has been operated commercially at Sasol I, II, and III. The naphtha requires reforming to meet fuel specifications.

The upgrading scheme for Synthol products, as practiced at Sasol II and III, is complex. Major processing steps include a heavy catalytic polymerization unit to upgrade C_3-C_4 olefins to gasoline and distillate, catalytic isomerization of the C_5-C_6 fraction, hydrogenation and reforming of the C_7-190°C fraction, hydrogenation and cracking of the 190°C+ fraction to lower its pour point, and aqueous-phase distillation and hydrogenation of oxygenates to convert aldehydes into alcohols. The Sasol Synthol process has been selected for the Mossel Bay natural gas conversion project in South Africa.

F-T Processes Under Development

Shell. Shell Oil Company, in 1985, announced its SMDS (Shell middle distillate synthesis) process, which produces primarily middle distillate. Shell recently announced plans to use this process to convert natural gas to distillate in Malaysia. The technology uses a fixed bed tubular reactor similar to the Arge design, with a more active catalyst believed to be a promoted cobalt catalyst. The syngas would be produced by a conventional partial oxidation process. The heavy, waxy product (mostly paraffins) is fed to a heavy-paraffin conversion reactor that uses a commercial Shell catalyst for hydroisomerization and hydrocracking to high-quality diesel. The naphtha product must be upgraded further.

Slurry F-T Reactor Design. The slurry reactor, first used by Fischer in the 1930s, was demonstrated at the Rheinpreussen-Koppers demonstration plant in 1953. In the late 1980s a major development was conducted at the Mobil Research and Development Corporation with partial funding by DOE. The development involved the upgrading of a total vaporous F-T reactor effluent over a ZSM-5 catalyst.

Other efforts in F-T technology include those by STATOIL, Gulf-Badger, Dow, Exxon, BP, Amoco, Mobil, and Union Carbide corporations.

Economics

The cost of syngas-based fuels, using reliable estimates, is between $45 and $62/barrel oil equivalent (10 percent discount rate; see Figure 3-1). Syngas-based fuels from natural gas are expensive because of the high value of domestic natural gas for conventional markets. For example, natural gas at $5/million Btu represents $33/barrel of the $60/barrel cost of MTG gasoline using the fluid bed reactor design.

Although Alaskan natural gas would be significantly less costly, higher capital and transportation costs for liquid fuels produced in Alaska would offset the gas cost advantage. For example, estimates show that a natural gas-to-methanol plant would cost 70 percent (California Fuel Methanol Study, 1989) more to construct at Prudhoe Bay than at a U.S. Gulf Coast location. Also, the cost of shipping methanol to Southern California from Prudhoe Bay is about $40/barrel oil equivalent ($22/actual barrel) compared to $7/barrel ($4/actual barrel) from the U.S. Gulf Coast. These costs assume that methanol shipped from Prudhoe Bay would require a new pipeline to Valdez and dedicated tankers from there to Southern California. Methanol from the Gulf Coast would require dedicated tankers. Because of methanol's lower energy density, greater volumes must also be shipped than would be the case for petroleum products.

Even if syngas-based fuels from natural gas were to become viable owing to a combination of cost reductions and special situations, exploitation would use foreign natural gas in a foreign location. At foreign locations, such as the Middle East, South America, or the Caribbean, natural gas would be significantly less costly than domestic gas because no local market exists and production costs are low. These foreign locations also meet the criteria of a reasonable construction cost environment and low transportation costs to major world markets. For example, estimates show that a methanol plant costs only 10 to 25 percent more in the Middle East and South American or Caribbean locations than in the U.S. Gulf Coast. Transporting methanol from these sites to Southern California was estimated to cost only $4/barrel oil equivalent (California Fuel Methanol Study, 1989).

Syngas-based fuels from coal are expensive primarily because of high capital charges for the large capital investment required. For example, at a 15 percent discounted cash flow rate of return, the capital charges for making gasoline using the fluid bed MTG design represents $43/barrel oil equivalent of the $78/barrel total cost.

The cost of syngas-based fuels from coal would be reduced if emerging technologies, such as the Shell gasifier and the slurry Fischer-Tropsch process, are used. Advanced gasifiers, such as the Shell gasifier, achieve a high thermal efficiency primarily by minimizing steam consumption. Syngas containing minimum amounts of hydrogen, i.e., having low hydrogen-to-carbon monoxide ratios, is produced. The slurry F-T process can use this syngas directly to make liquid fuels without further addition of steam or water; thus, maintaining high overall efficiency.

Although advances in technology will reduce the cost of syngas-based fuels, it is unlikely that these cost reductions will reduce costs to $30/barrel primarily because the technologies are relatively mature. According to Schulman and Biasca (1989), the cost of methanol from coal would be reduced by $4/barrel oil equivalent due to improved synthesis gas unit designs. Using western coal in a fluid bed gasifier would reduce the cost of methanol by an additional $6/barrel.

Methanol from underground coal gasification (UCG) costs less than from surface gasification; that is, $44 to $57/barrel rather than $53 to $65/barrel. However, since the reliability of the estimate is considered speculative, conclusions about UCG cannot be made at this time.

Conclusion and Recommendations for the DOE Program

Processes to produce methanol or F-T liquids from syngas continue to be studied vigorously by industry, and methanol and F-T liquids may well find application in the United States. Production is, however, expected to be primarily outside the United States, where low-cost natural gas is available. While an interesting area of research in which further advances can be expected, the above factors discourage DOE research beyond that of fundamental and exploratory research.

DIRECT COAL LIQUEFACTION

U.S. recoverable coal reserves are large, representing a significant proportion of world energy resources, and their prices are likely to remain modest (EIA, 1989a,b; Table 1-2; Table D-2). Unless a major new coal conversion industry is developed very rapidly, any shortage of coal is unlikely, the industry growth should not strain U.S. engineering or construction capabilities, and the price advantage of coal as a feedstock over natural gas and petroleum will probably improve.

Coal has additional advantages over other solid feedstocks that might be converted to liquid fuels. Because coal is geographically dispersed, its commercial use will produce less concentrated environmental impacts and more manageable demands on local infrastructure. Coal also has a high concentration of hydrocarbons, reducing mining, transportation, and processing costs. There is also a well-established U.S. mining industry.

Technology

In direct liquefaction, hydrogen is added to coal in a solvent slurry at elevated temperatures and pressures. The process was invented by Freidrich Bergius in 1913 and was commercialized in Germany and England in time to provide liquid fuels during World War II. The first U.S. testing of direct liquefaction processes followed World War II (Kastens et al., 1949); efforts in the area declined when inexpensive petroleum from the Middle East became available in the early 1950s. Interest revived when the Arab oil embargo of 1973 caused high oil prices, resulting in increased federal funding for such research. A variety of process concepts were examined on a small scale, and three were tested on a large scale in the late 1970s and early 1980s: SRC-II (solvent refined coal) in Tacoma, Washington; EDS

(Exxon donor solvent) in Baytown, Texas; and H-coal in Catlettsburg, Kentucky. The DOE provided most of the funding for these successful demonstrations, but none proved economical as oil prices fell in the early 1980s.

U.S. research continued after the big pilot plants were abandoned, most of it funded totally or in part by DOE. Few of the smaller pilot plants survived; today the only integrated pilot plant operating full time on direct coal liquefaction is the Advanced Coal Liquefaction R&D Facility in Wilsonville, Alabama. Test units are available for contract at Hydrocarbon Research, Inc., Lummus-Crest, and the University of Kentucky, and Amoco Oil Company has smaller bench-scale pilot plants in operation.

As indicated in this report, the projected cost for making liquid fuels from coal is about $40/barrel. This is about half of the approximately $60 to $80/barrel projected 10 years ago (Lumpkin, 1988). Although many contributed to improving the technology, the Wilsonville pilot plant has become the focus for the U.S. direct coal liquefaction program, and estimates of the improved economics are based on the technology demonstrated there. The process at Wilsonville initially employed a single-stage dissolver, followed by a filter to remove undissolved coal solids and a still to recover the solvent for reuse in the dissolver. The product was a solid with low ash and sulfur content to be used as boiler feed. The facility evolved in steps into a more versatile 6-ton/day liquefaction plant.

The improvement in economics cannot be attributed to any single breakthrough but is rather the accumulation of improvements over several years of operation. First, a more effective and reliable process to remove solids from the liquid product by controlled precipitation replaced the filter. A second catalytic reactor was added to improve control over the chemistry of liquefaction. This reactor was first installed downstream of the solids removal and distillation systems; moving the reactor upstream further improved operation. Some of the recycled liquid used to slurry the feed coal was then bypassed around the solids removal unit, increasing the efficiency of the unit. Improved catalysts were added to both the first and second reactors. This series of modifications led to higher liquid yields, improved conversion of nondistillable liquids, less rejection of energy along with discarded coal minerals, and increased throughput relative to early two-stage systems. The success of this evolution shows that steady, long-range R&D can achieve major technological advances.

The U.S. direct liquefaction process appears superior, but there is significant related activity overseas. The Japanese are operating a 50-ton/day liquefaction plant in Australia (Coal and Synfuels Technology, 1989b). Information on its operation is sparse, but apparently there have been some difficulties. The Japanese say they have developed a catalyst superior to any on the market, but samples are not available for testing. A 150-ton/day pilot plant, using a similar process, is being designed for installation in the

1990s in Japan. The West Germans have operated a 200-ton/day pilot plant at Bottrop since the early 1980s, but it was recently converted to the study of upgrading petroleum residuum. The West Germans have an unmatched backlog of experience in coal liquefaction on a large scale, although their process has some drawbacks compared to recent U.S. developments. The British are building a pilot plant about the size of Wilsonville in North Wales, at Point of Ayr. Their smaller-scale work was very encouraging, and this process may be a strong competitor if it works out on the larger scale. All of these projects are government funded.

Fuel Properties

Products of direct coal liquefaction are expected to meet all current specifications for transportation fuels derived from petroleum. Major products are likely to be gasoline, propane, and butane. Distillate fuels can be made but would likely require large volumes of hydrogen. Gasoline may be a particularly attractive product because it would have a relatively high octane.

High octane is achieved by the high aromatic content of the liquids. If regulations are established limiting the aromatic content of gasoline for environmental reasons, the cost of liquid fuels produced from coal by direct liquefaction would rise. While the benzene content of gasoline made from coal is extremely low, the concentration of other aromatics is high, and they could be hydrogenated to produce naphthenes at a moderate increase in cost. This would increase the volume of the products, decrease octane number, and increase hydrogen consumption.

Environmental Considerations

With regard to environmental emissions with local impact, a coal liquefaction facility is broadly comparable to a refinery with up-to-date emission control systems. As in many conversion processes, sulfur and nitrogen are removed from the feedstock and appear in the product fuels at greatly reduced levels. Wilsonville has successfully shown that local emissions can be controlled satisfactorily (see Chapter 5 regarding greenhouse gas emissions). Direct coal liquefaction is fairly energy efficient; about two-thirds of the coal fed comes out as liquid product, and the rest is consumed to run the process.

One area of concern is industrial hygiene. The intermediate products of coal liquefaction (internal to the process plant) are polynuclear aromatic hydrocarbons, which are well-known carcinogens and mutagens. Industry and government programs over the past 20 years have demonstrated that proper attention to hygiene can make coal liquefaction plants safe places to work (U.S. DOE, 1989b).

Potential Cost Reduction

The improvements demonstrated at Wilsonville were incorporated in a new design that is an update of the DOE-funded Brechenridge Project, which used the H-coal process and was completed in 1981. The new design uses two reactors in series in place of the single-stage H-coal design and includes a simplified product distillation system, the new solids removal technology, and the simpler water treatment methods proved at Wilsonville; the process upgrades most of the liquid products to gasoline blending components (Lumpkin, 1988). The result is a projected cost decrease of about 60 percent, to about $40/barrel (crude oil equivalent).

Substantial improvements are still likely if research continues. Recent runs at Wilsonville have achieved a doubling of coal feed rates from that assumed in the above design by making the reactor temperatures more uniform. These high rates may reduce costs by $2 to $3/barrel. Equipment is being installed to allow better distillation of the product liquids, which will in turn reduce the amount of high boiling components in the products. This change may reduce costs another $2/barrel.

There is a large incentive to learn how to process different coals. Most recent Wilsonville trials used Illinois coal. Runs with a higher-rank, bituminous Ohio coal have demonstrated very high liquid yields, but the quality of the liquids needs to be improved. Lower-rank coals, either subbituminous or lignites, are relatively cheap and convert to high-quality liquids. However, their high moisture content and high levels of oxygen lead to problems that have not been entirely resolved at Wilsonville. Resolution of these problems could reduce costs by as much as $4/barrel.

Removal of ash, and perhaps the unreactive parts of the coal itself, before liquefaction could improve conversion, reduce erosion, and eliminate the need for the current solids separation process. Technologies are under development, funded primarily through DOE or the Electric Power Research Institute, that might be used to clean liquefaction feedstock, although they are primarily intended to prepare cleaner power plant fuel. Capital investment at Wilsonville is needed to adapt these technologies and determine their economic attractiveness.

There are clearly many opportunities to improve the economics of direct coal liquefaction. The DOE hopes to reduce costs at Wilsonville by 15 percent within the next 3 or 4 years. This target seems conservative.

Direct liquefaction is capital intensive, and its total cost is relatively insensitive to most individual improvements. Multiple improvements are needed to significantly reduce costs. Such improvements are likely if research continues.

Environmental requirements to reduce the aromatic content of gasoline may increase the costs of producing this fuel from coal.

Although it is impossible to predict whether major technical breakthroughs will occur, the possibility should not be discounted. The high level of U.S. and foreign fundamental research on coal structure and chemistry for the past 10 years could lead to a superior means of conversion.

DOE Program on Direct Liquefaction

The recent assessment of research needs conducted by DOE's Office of Program Analysis outlines an excellent program aimed at bringing down the cost of direct liquefaction (Schindler, 1989). Industry members of the assessment panel particularly stressed the need for federal funding of a large-scale pilot plant, as large as the German effort at Bottrop or the planned Japanese pilot plant (i.e., processing 150 tons per day or greater) to develop hardware and perform chemical studies. A broad range of fundamental and exploratory research was also recommended, based on the recognition that possible improvements in the current technology appear limited but that breakthroughs may bring down the cost of liquid fuels produced from coal to below $20/barrel. The committee concurs with these research recommendations.

In between these two extreme types of development, intermediate-size flow units are needed. Wilsonville, or an alternative plant of about the same size, would be useful to test changes in process configuration at reasonable cost. Smaller pilot plants are also needed to test catalysts, explore operating conditions, and provide low-cost testing of new ideas. The committee recommends that such small-scale work continue to be sponsored by DOE, with the work performed by private contractors in industry and universities. In this way a wide variety of experts can contribute and technology transfer to industry will be enhanced. The general purpose pilot plant proposed for installation at the Pittsburgh Energy Technology Center is less attractive due to lower industrial participation.

The DOE-funded programs that are relevant to the conversion of coal into transportation fuels in fiscal year 1990 allocate approximately twice as much money to process development as to each of the other categories (see Appendix G for definitions of fundamental, exploratory and catalyst, and process research). This emphasis on development may be unavoidable when industry is reluctant to participate because of the long time scale and uncertainty involved. The accuracy of the funding split is somewhat uncertain, since it is based on brief project descriptions that DOE provided the committee. However, process demonstration, which is the step following process development, is receiving no funding, and without money for this purpose over the long run the United States will fall behind its foreign competitors.

When the pilot scale demonstrates that processing 150 tons or more of coal per day has provided scale-up information, proceeding with a commercial-scale demonstration facility should be seriously considered. This facility should be a single-line plant with enough redundancy of critical equipment that an acceptable plant on-stream factor can be achieved. Only by operating a commercial-scale facility can its true economics be determined.

It would be best if this commercial-scale demonstration could be financed on an international scale. Countries such as West Germany, France, Italy, and Japan depend on imported oil for transportation fuel. An international project funded by their governments and the United States on a cost-sharing basis, with participation of the private sector in the form of project management, engineering, and construction, operations, and maintenance, as well as private sector investment, would help ensure a successful effort. The Research, Development, and Demonstration organization of the International Energy Agency might coordinate and monitor the project. A successful commercial-scale demonstration would be valuable for the United States should it become desirable for the supply of liquid transportation fuels to be augmented through direct coal liquefaction technology.

Conclusion

Over the next 5 years research effort directed toward new catalysts and new processes should be stressed with a goal to selecting the best coal conversion processes for demonstration in a large pilot plant within this time frame. Achievement of this goal will require establishing technical confidence, achieving anticipated environmental requirements, and reducing the cost so that industry is willing to participate. The program should include high-quality economic and technical evaluations by engineering firms, petroleum industry operating companies, and qualified consultants to guide the selections of the best technologies to move forward.

COAL-OIL COPROCESSING

Coal-oil coprocessing is a technology that simultaneously converts heavy petroleum residuum and coal to liquid transportation fuels. Incentives for coprocessing depend strongly on the existence of synergisms between the coal and resid as they are processed together. Coal may aid operability due to the solvency of coal liquids, and coal ash may scavenge metals from the resid to extend catalyst life. Other synergies may exist. In the 1970s the Canada Centre for Mineral and Energy Technology (CANMET) showed that the addition of less than 5 percent coal to a petroleum feedstock significantly improved distillate product yields (Rahimi et al., 1987). This process was employed in a 5000-bbl/day plant started up in 1985 by Petro-Canada, near Montreal (Kelly and Fonda, 1984). Background, state-of-the-art, and R&D opportunities for coprocessing technology are summarized in two recent reports (Schindler, 1989; Schulman et al., 1988).

The most significant disadvantage of using coal in petroleum upgrading is its impact on capital and operating costs—coal handling, hydrogen generation, catalyst replacement, and waste amelioration.

Recent Developments

Various coprocessing technologies have undergone some development. Chevron Corporation ran a 6-ton/day coprocessing pilot plant at its Richmond, California, facility in 1983 (Shinn et al., 1984). Results reported for this close-coupled thermal catalytic system included good operability, a synergy for resid conversion, and demetallation of high-metal-content resids. In 1984 Kerr-McGee tested a process in which the bottoms from a resid hydrotreater replace about one-half the recycle oil in a thermal catalytic two-stage coal liquefaction process (Rhodes, 1985). Lummus coprocessing technology includes two process flow schemes (Greene et al., 1986). In one, coal and hydrotreated resid are fed to a two-stage process consisting of a short contact time (SCT) thermal reactor and an expanded bed LC-fining system. In the second scheme the resid is fed to the LC-finer only. In either case a solvent stream is recycled to the first stage, the product vacuum bottoms are fluid coked, and the coke is gasified.

Significant foreign developments have occurred in the West German 250-ton/day pilot plant operated by Veba Oil at Bottrop (Schulman et al., 1988). It was used as a coal liquefaction pilot plant until 1986 and thereafter processed petroleum vacuum resids. Although not used for coprocessing, its operation with both coal and petroleum resid indicates the flexibility of the technology to accommodate different feedstocks.

Current Developments

The DOE is supporting the development of coprocessing in two pilot-scale programs at Hydrocarbon Research, Inc. (HRI), and UOP, Inc., and in various smaller-scale research projects. In addition, under the first round of DOE's clean coal technology program the department selected a 12,000-bbl/day coprocessing project, using HRI technology, sited in Ohio. If this project is completed, it could mark the first large-scale demonstration of coprocessing technology. Further descriptions of the Signal-UOP and HRI technologies are given in Appendix H.

In addition to these pilot-scale projects, advanced and fundamental coprocessing research is being conducted at U.S. universities and research institutions. Studies concern coal-oil interactions and process chemistry, with the goal of improving existing processes and identifying new process concepts. Some research on direct liquefaction and heavy resid upgrading may also apply to coprocessing.

Fuel Properties

Coprocessing is similar to direct coal liquefaction in that it produces fuels that are compatible with existing fuel markets. In particular, it is directed toward producing transportation fuels because these are the highest value-added products. Therefore, the emphasis has been on producing refinery feedstocks and finished products that can meet motor and jet fuel specifications. The mid-distillate and vacuum gas oil products from coprocessing are low in sulfur and nitrogen (Duddy et al., 1986) and could be used as low-sulfur fuel oil or turbine fuel for utility applications.

Environmental Considerations

The aromatic nature of coal tends to impart aromatic content to the products, which improves the octane value of the product naphtha for use as a gasoline-blending stock.

Burning fuels from coprocessing will emit no more pollutants than their petroleum-derived counterparts. However, all fuels are produced at some loss in thermal efficiency and CO_2 is produced. Other environmental impacts of coprocessing should be within the scope of existing petroleum refining and coal utilization practices.

Opportunities for Cost Reduction

Determining the existence and extent of synergism between coal and resid is needed to assess the economics. Several investigations place the product costs of coprocessing between those of heavy resid upgrading and direct coal liquefaction (Schindler, 1989; Schulman et al., 1988; Duddy et al., 1986; Huber et al., 1986). Cost reduction will come from determining how to maximize the benefits of any synergisms that might exist. In particular cases a combination of appropriate refinery equipment, resid costs, and coal availability might justify coprocessing.

It appears unlikely that coprocessing will find application as a stand-alone technology. The economics of coprocessing require a significant difference between coal and resid costs to justify the additional capital expense to add coal to an existing refinery. Such a gap between coal and petroleum prices could justify construction of a coal liquefaction plant rather than a grass-roots coprocessing plant, to take greater advantage of coal as the less expensive feedstock. Of course, a coal liquefaction plant would probably be capable of coprocessing or resid hydrocracking if resid costs decrease.

Other opportunities for cost reduction in coprocessing are the same as those in direct coal liquefaction: increased throughput, reduced hydrogen cost, lower coal costs, better catalysts, better engineering design, and identification of new process concepts.

Recommendation for the DOE Program

Coprocessing of heavy oils or residuum with coal may offer an opportunity for the introduction of coal as a refinery feedstock. A demonstration plant for production of a clean boiler fuel is part of the DOE's clean coal technology program. Funding of basic bench-scale research should be continued over the next 5 years to define the extent of synergism for coprocessing coal-resid combinations, followed by a thorough economic analysis quantifying the impact of this synergism. If favorable, the impact of synergism should be confirmed at the Wilsonville test facility to define optimum processing conditions. If little or no synergism is found, work in this area should be terminated.

COAL PYROLYSIS

Description of the Technology

Pyrolysis of coal dates back to the 18th century, using temperatures below 700°C in fixed or moving bed reactors. The primary product was a low-volatile smokeless domestic fuel, although the value of the liquid products was also soon recognized. During the 1920s and 1930s there was a great deal of R&D in low-temperature processes, but interest died in the mid-1940s when gas and oil became readily available at low prices. With the oil embargo and increased oil prices of the early 1970s, interest renewed in coal pyrolysis, but in more recent times interest has again declined along with petroleum prices (Khan and Kurata, 1985).

In the most recent work, development was aimed at processes that maximize the yields of liquid products. These processes require rapid heat-up, using fluidized or entrained bed reactors. A number of the processes require the addition of reactants (steam, carbon dioxide, and hydrogen) at greater than atmospheric pressure to increase yields and limit secondary reactions. Reactor type, temperature, pressure, residence time, and coal type all have significant impacts on product yields (see Appendix I, Tables I-1 and I-2).

Pyrolysis under mild temperatures (500° to 700°C) and pressures (up to 50 psig) with rapid heat-up can produce high liquid yields without adding hydrogen (hydrogen would have to be added to these liquids to produce transportation fuels). However, a significant part of the feed coal remains as char with less market value than the feed coal. As a result, coal pyrolysis offers the promise of lower liquid costs only if the char can be upgraded to a higher-value product, such as form coke, smokeless fuel, activated carbon, or electrode carbon, or if the liquid yield can be significantly increased by using low-cost reactants (steam and carbon dioxide) or catalysts.

Fuel Properties

Different liquid fuel properties of the products of coal pyrolysis result from different processes. Processes include the Coalite process, a slow heating process that produces more gases and char than tar (Khan and Kurata, 1985); the Occidental Research Corporation flash pyrolysis process, a rapid heat-up process that produces more tars (DeSlate, 1984); and the FMC COED process (see Appendix I for properties). For the above order of processes the desirability of the char as a fuel decreases because of the increase in ash and sulfur content on a heat content basis and the deterioration of its size consistency, which can lead to material handling problems.

Pyrolysis liquids require extensive hydrogenation to be useful as transportation fuels. Another approach is to combine coal pyrolysis with production of synthesis gas to potentially increase the liquid yields for conversion processes producing transporation fuels.

Environmental Considerations

The environmental impacts of using liquid and char coproducts from coal pyrolysis will be very similar to those associated with the feed coal. Coal pyrolysis in the presence of alkaline material can result in the retention of sulfur in the char with a corresponding reduction in the liquid products. Care must then be taken to avoid uncontrolled releases of hydrogen sulfide from the char alkaline mixture (Gessner et al., 1988).

Economics

The timing of a commercial application of mild pyrolysis will depend on the marketability of the char and the quality of the liquids. Spot market prices for metallurgical coke now exceed $130/ton. Assuming that form coke can command a comparable cost, mild pyrolysis could be economically viable in the immediate future.

Recommendations for DOE Research

To realize the potential of mild pyrolysis as a source of transportation fuels, a number of research issues must be resolved. The DOE's current research program partly addresses these issues. Under its Surface Coal Gasification Program, research is focused on the development of advanced continuous mild gasification systems to produce optimal readily usable coproducts. The program is funding the development of four 100-lb/hour pilot units using fundamentally different approaches.

The DOE should continue funding this program, with special attention to the quality of the liquids produced, the value and marketability of the char, and the size of the coal resource base that can be used with this technique. In the future DOE should attempt to scale up one or a combination of the most promising processes to obtain adequate design data for a commercial demonstration.

The committee concurs with a number of recommendations on coal pyrolysis R&D made by an assessment panel to DOE (Schindler, 1989). The first was to study the chemistry and mechanisms of catalytic hydropyrolysis. Another important recommendation was to conduct a systems analysis of pyrolysis or hydropyrolysis coupled with gasification and combustion as a means of utilizing the char.

DIRECT CONVERSION OF NATURAL GAS

Recently, substantial research activities have been conducted in the area of natural gas (methane) conversion to methanol without the use of syngas (Kuo, 1984; Kuo et al., 1987). The primary commercial goal of this research is to convert remote natural gas, which cannot easily be brought to market and is of little commercial value, into more easily transported liquid fuels. Except for Alaska, all of the remote low-cost gas is located in other countries.

Technology and State of Development

Numerous direct methane-to-methanol conversion routes are being studied at the bench scale by various companies, government agencies, and universities. These include cold flame oxidation (direct partial oxidation) in which the main chemical reaction is the oxidation of methane to methanol, direct oxidation involving the catalytic coupling of methane and an oxidant to produce C-2 products and hydrocarbons, oxychlorination, indirect oxidation with oxidative coupling to ethylene, and catalytic pyrolysis involving contact of methane with a catalyst. ARCO Oil and Gas Company appears to be a leader in indirect oxidation, and recent success has been reported with its REDOX process. Other conversion routes include strong acid conversion and biological conversion (see Appendix J for additional technical details).

Economics

Although the technology exists for converting natural gas into liquid fuels, the cost is too high when the technology involves first converting natural gas into syngas. The costs for syngas-based fuels from natural gas, expressed as 1988 dollars per crude oil equivalent barrel, depends on the technology and assumed rate of return and vary as follows: (1) methanol from natural gas, $45 to $50/barrel; (2) methanol to gasoline, $60 to 68/barrel; and (3) the Shell Middle Distillate Synthesis process, $58 to $64/barrel (see Chapter 3 and Tables D-3 and D-4).

The direct conversion routes have the potential of being more energy efficient and less expensive since they bypass the formation of syngas. However, the current level of development has not achieved the potential significant cost reductions. Gasoline from ARCO's REDOX process costs more than Mobil's MTG (fluidized bed) process (Schumacher, 1989).

An analysis of the cold flame oxidation route, showed that based on an optimistic design the cost of gasoline would be reduced only 7 to 15 percent (with zero cost for natural gas) compared to the conventional MTG technology (Fluor Corporation, 1988). This analysis also indicated that the cold flame oxidation route did not have any overall thermal efficiency advantage.

Liquid fuels from domestic natural gas are expensive because of the high value of domestic natural gas for conventional markets. For example, natural gas at $5/million Btu represents $33/barrel of the $60/barrel (10 percent discount rate) cost of MTG gasoline using the fluid bed reactor design. Even if a direct methane conversion process were developed that used 20 percent less natural gas, the cost of natural gas would represent $26/barrel of crude oil equivalent of the gasoline cost.

Although Alaskan natural gas would be significantly less costly, higher capital and transportation costs for liquid fuels produced in Alaska would offset the gas cost advantage. Estimates show that a natural gas-to-methanol plant would cost 70 percent more to construct at Prudhoe Bay than at a U.S. Gulf Coast location. Also, shipping methanol to Southern California would cost about $40/barrel oil equivalent from Prudhoe Bay compared to $7/barrel from the U.S. Gulf Coast (California Fuel Methanol Study, 1989).

Even if liquid fuels from natural gas were to become viable owing to a combination of cost reductions and special situations, exploitation would use foreign natural gas.

At foreign locations, such as in the Middle East, South America, and the Caribbean, natural gas would be significantly less costly than domestic U.S. gas because no local market exists and production costs are low. These foreign locations also meet the criteria of a reasonable construction cost environment and low transportation costs to major world markets. For example, methanol plant costs are only 10 to 25 percent more in Middle Eastern, South American, and Caribbean locations than on the U.S. Gulf Coast. Transporting methanol from these sites to Southern California was estimated to cost only $4/barrel oil equivalent (California Fuel Methanol Study, 1989).

Recommendations for the DOE

Numerous direct methane conversion routes are being studied at the bench scale by various companies, government agencies, and universities that avoid the need to produce syngas as an intermediate. These direct conversion routes have the potential of being more energy efficient and less expensive since they bypass the energy-intensive and expensive step—the formation of syngas. However, the current level of development has not achieved the potential significant cost reductions.

Even if liquid fuels from natural gas were to become viable owing to a combination of cost reductions and special situations, exploitation would use less valuable foreign natural gas in a remote location. Therefore, government-sponsored research on direct methane conversion technology should be limited to fundamental research.

5

Environmental Impacts of Alternative Fuels

AIR QUALITY, HEALTH, AND SAFETY EFFECTS

The activity involved in producing, manufacturing, distributing, and using fuels raises issues on a variety of adverse effects on the environment and on human health and safety. Environmental effects from liquid fuel production are presented in Chapter 4; only the end-use effect on the environment of using these alternative fuels is considered here (greenhouse gases are considered later). Pollutants, emissions, air quality, safety, and toxicity are addressed. Economic aspects of alternative fuels and vehicles are considered in Chapter 3.

Currently, U.S. local, state, and federal governments are interested in reducing emissions from motor vehicles and stationary sources in those regions that fail to meet ambient ozone standards. One proposed strategy to reduce ozone is to decrease emissions from gasoline-powered vehicles of volatile organic carbon (VOC) compounds and nitrogen oxides (NO_x), precursors to ozone formation. The California Air Resources Board (CARB) has established a hydrocarbon exhaust standard about 40 percent more stringent than the U.S. Environmental Protection Agency's (EPA) standard of 0.41 g/mile. Clean Air Act bills have been introduced to require a similar standard for the other 49 states. CARB is also considering an even more stringent emission standard for vehicles, as required by the air quality management plan for the Los Angeles area. Efforts are also under way to design engines or use fuels that will meet the stricter diesel particulate standards in the 1990s. The U.S. government is encouraging the use of alternative fuels through the testing of demonstration fleets and research on alternative-fueled vehicles (U.S. Congress, 1988b). In addition, President

Bush's Clean Air Act proposals call for the introduction of such vehicles in urban areas that have unusually severe air quality problems.

Important vehicle fuel options include compressed natural gas (CNG), methanol, and reformulated hydrocarbon-based fuels. Electric vehicles could potentially be used in niche markets and could certainly reduce automotive emissions (Wang et al., 1989). Hydrogen vehicles are a potential long-term option (DeLuchi, 1988; Ogden and Williams, 1989). Reformulated fuels, electricity, and hydrogen are not considered here in any detail. Analysis of alternative fuels and emissions requires the consideration of which fuel-engine-emission control technology combinations are capable of meeting emission standards with the least cost and inconvenience to motorists. These trade-offs are not considered in detail in this study.

Automobile Exhaust Emissions and Air Quality

There is a great deal of uncertainty in determining the impact on air quality of different fuels. A variety of air quality impacts between different fuels have been reported, especially for ozone (Alson et al., 1989; DeLuchi et al., 1988b; Dunker, 1989; Harris et al., 1988; Long et al., 1986; Moses and Saricks, 1987; Nichols and Norbeck, 1985; Sierra Research, 1988; Whitten et al., 1986). First, actual emission rates are determined by trade-offs between emissions standards, costs, performance, and driveability. If a particular fuel offers the potential for easier emission control, then engines designed to emit the maximum allowed can gain other benefits such as reducing the cost and complexity of pollution control equipment and increased performance. Actual emissions will likely vary considerably across vehicle make and model. Also, there are limited emission data for alternative-fueled vehicles at low mileage and virtually no data on performance of their emission control systems at high mileage and in actual use by typical motorists.

Most pollutant production is sensitive to the air-to-fuel ratio of engines. If future engines are designed to run "lean" (using high air-to-fuel ratios) to achieve greater fuel efficiency, then for moderately lean mixtures the NO_x levels would be higher and carbon monoxide and hydrocarbon emissions and engine power would be lower than those of engines operating at stoichiometric ratios, as do most of today's gasoline engines. However, almost all automobiles now (and for the foreseeable future) use three-way catalysts with the air-to-fuel ratio controlled to approximately stoichiometric conditions. This means that comparisons of the carbon monoxide production at lean fuel mixtures are valid but are irrelevant unless the catalysts and air-to-fuel ratios used in automobiles are changed. R&D is being conducted on lean-burn engines, but to date they have not been able to meet emission standards for oxides of nitrogen.

A distinction must also be made between single-fuel fully optimized engines and multifuel engines, and the fuel used must be specified clearly since some methanol emission data are based on a fuel consisting of 100 percent methanol, while others contain 10 or 15 percent gasoline. Analysis becomes even more complicated for multifuel methanol-gasoline engines, since they will operate on various blends of methanol and gasoline. For improved cold starting and flame visibility, about 15 percent gasoline will probably be added to methanol. Moreover, specifying environmental effects is complicated. The ozone formation process is highly complex; even the most sophisticated photochemical air quality models have error margins of 30 percent or more in predicting hourly averaged ozone concentrations (Russell, 1988; Tesche, 1988). Ozone formation also depends on the ratio of reactive hydrocarbons to oxides of nitrogen (RHCs/NO_x) in the atmosphere, which varies greatly among urban areas. Only in the Los Angeles area have sufficient meteorological and spatial pollutant concentration data been collected to operate multiday photochemical airshed models; results from Los Angeles, however, cannot be generalized to other regions.

Surveys of published data generally conclude that a reduction in peak ozone level of 0 to 20 percent might be attainable from complete substitution of methanol for gasoline (Beyaert et al., 1989; DeLuchi et al., 1988b). Generally, more recent studies using more realistic assumptions predict less benefit than older studies. Even the recent studies assume catalytic reduction of formaldehyde to low levels, which has not yet been demonstrated at high mileage or for public use. Emissions of formaldehyde, which is photochemically very reactive, are about three to six times higher from current methanol-fueled prototype cars than from those operated on gasoline, and technology to bring formaldehyde emissions down to gasoline fuel levels has not yet been demonstrated, though California has emission standards in place requiring major reductions before methanol vehicles can be sold, beginning in 1993.

While not yet clearly established, it appears that, if the formaldehyde concentration in the methanol exhaust can be reduced to that produced by gasoline-fueled cars, some regions may reduce smog by using methanol. Present catalysts, optimized for gasoline, do not attain this goal. Whether new catalysts can (at high mileage in actual use) achieve the levels set by California is not yet clear. The problem is further complicated when multifuel cars are used—it is more difficult to find a catalyst equally effective for both fuels.

CNG vehicles emit very little carbon monoxide (if operated at a lean air-to-fuel ratio) and much less reactive hydrocarbons than gasoline vehicles. CNG will have less smog-producing reactive exhaust pollutants than gasoline vehicles if NO_x emissions can be controlled adequately. Most data to date indicate that CNG vehicles may produce as much, or more, NO_x because the current three-way catalyst technology is not effective unless carbon monoxide is present. It is too soon to have any air quality models of the impact of different reformulated gasolines on airsheds.

In summary, there is considerable uncertainty about the air quality benefits of practical alternative fuel-engine-emission control combinations over present fuels. Methanol produces more formaldehydes in the exhaust, which can react immediately to produce smog. This will require development of effective and durable emissions controls. Methanol's ability to reduce smog depends on solving the cold-starting and engine durability problems of methanol vehicles, the development of catalysts more effective at reducing aldehyde emissions, and nontransportation factors that affect the RHC/NO_x ratio in the atmosphere in different airsheds. Reformulated gasoline could also be used in all vehicles and may also produce air quality benefits without the cost, consumer acceptance, and other problems associated with establishing a new fuel distribution system and redesigning vehicles, but no data are available to support this. Additional R&D is needed before valid judgments can be made about the comparative environmental effects of various fuels.

Air Quality Impacts of Diesel Engines

The U.S. diesel fuel market is about 0.58 billion bbl/year compared to 2.6 billion bbl/year of gasoline use for 1988. CNG and methanol may be used in modified compression ignition (diesel) engines. The conclusions on spark ignition engines are roughly the same for compression ignition engines. Use of methanol or CNG instead of diesel fuel dramatically reduces particulates emissions and, because there is no sulfur in these fuels, sulfur oxide emissions; methanol may also significantly reduce NO_x emissions depending on the engine design (Alson et al., 1989; Unnasch et al., 1986). However, heavy-duty methanol engines emit substantially more methanol, formaldehyde, and carbon monoxide than do diesel engines and would require oxidation catalyst systems and evaporative emissions controls that are not now needed for diesel engines. It is believed that satisfactory catalytic oxidation systems cannot currently be designed for the wide range of exhaust temperature and composition encountered in diesel exhaust. However, the continued use of diesel fuel will probably also require expensive changes and additional costs to meet 1991 and 1994 emission standards.

Because of EPA emission regulations that take effect in 1991, diesel urban transit buses will probably be the first market penetrated by methanol and CNG, but this market is dispersed and small, representing a total of only about 30,000 bbl/day in the United States (ORNL, 1987). Further penetration of the diesel market is likely to lag behind penetration of the gasoline market, because of high fuel costs and poor compression ignition characteristics and because diesel engines are not replaced as often.

Safety

There are different safety issues for the different alternative fuels. For example, leakage of natural gas from pressurized tanks in closed spaces creates the potential for explosions. Pure methanol burns with an invisible flame in daylight, although adding 15 percent gasoline makes the flame visible. Under normal ambient temperatures, methanol produces a flammable mixture in storage tanks, vapor control systems, and vehicle fuel tanks. Although there are different fire and safety issues for different fuels, proper engineering, handling, and education may adequately address these differences.

Toxicity

Methanol is toxic, as is gasoline. Ingestion of methanol is followed by a 12- to 24-hour latent period of no symptoms and then signs of poisoning, progressing to possible blindness, permanent neurological damage, and death without prompt medical attention. Since gasoline itself is toxic, combining gasoline and methanol raises questions about the toxicity of, and medical treatment for exposure to, the blends compared to each fuel alone (Beyaert et al., 1989). There are also concerns about chronic toxicity from human absorption through skin or inhalation in atmospheres with low ambient concentrations of methanol.

Drinking water can be contaminated by either methanol or gasoline, but odor and taste thresholds for methanol are much higher and it may not be detected in water supplies until significant human exposure has occurred. Methanol also has a significant cosolvent effect on water-insoluble hydrocarbons, possibly leading to increased concentrations of hydrocarbons in groundwater from releases of methanol and gasoline mixtures. Better understanding is needed of the effects of exposure to methanol liquids and methanol and formaldehyde vapors.

There are also concerns about the long-term effects of exposure to hydrocarbon vapors, particularly benzene and the other volatile aromatics. At an appreciable cost, aromatics can be converted to naphthenes by hydrogenation; however, the elimination of tetraethyl lead has made current gasolines and vehicles more dependent on the high octane number of aromatics. This removal would require replacement by other high-octane-number components such as oxygenated hydrocarbons at some increased cost and with uncertain impacts on air quality. An alternative is use of low-compression-ratio engines that will tolerate low octane number. These engines would suffer from higher fuel consumption for the same power, since efficiency is reduced by reducing compression ratios. Another alternative is development of stratified charge engines, which are less sensitive to octane number. Except for diesel engines, stratified charge engines, despite considerable development effort, have not yet been successfully commercialized.

Conclusions and Recommendations

Because of different assumptions about vehicle emissions, vehicle replacement rates, and ratios of RHCs to NO_x ambient levels, air quality models incorporating substitution of gasoline-powered vehicles by methanol-fueled vehicles vary widely in predictions of tropospheric ozone reduction. Using optimistic assumptions for the California South Coast Air Basin and some other areas, significant potential improvements are predicted if formaldehyde emissions can be controlled adequately. These results are very sensitive to modeling uncertainties and to the ratio of RHCs to NO_x in the ambient atmosphere, and they depend heavily on developing more effective catalysts and other control technology. Actual effects may range from adverse to beneficial. Ozone benefits of CNG have received less research attention but are likely to be greater than those of methanol if NO_x emissions can be controlled. Impacts of reformulated gasolines have not been extensively investigated, but ARCO recently introduced a reformulated regular gasoline in Los Angeles that is claimed to be significantly less polluting.

Methanol has different health and safety impacts than gasoline, but it is not necessarily superior or inferior to gasoline. CNG is not toxic and may be generally safer than gasoline and methanol. The committee makes the following recommendations:

• The DOE should cooperate with the automobile and fuel industries and other government agencies such as the EPA and the National Institutes of Health (NIH) to investigate the opportunities for reducing automobile emissions with investigations using a total systems approach—resource to disposal—to adequately compare fuel-engine-emissions control combinations. This research would facilitate the design and production of cost-effective, environmentally acceptable fuel-vehicle combinations.

• DOE, along with EPA and others, should try to resolve uncertainties about the air quality and health effects of fuel and vehicle options. Emissions quantities and compositions for advanced-technology vehicles need to be evaluated more extensively, simulating use by typical motorists and employing more accurate photochemical air quality models for different urban areas. This work should lead to a comprehensive data base, independent of any decisions (government or private) to commercialize particular engines or fuel types.

GREENHOUSE GAS EMISSIONS

As mentioned previously, the accumulation of greenhouse gases in the atmosphere may lead to global warming. The extent of global warming, its timing, and its potential impacts are highly uncertain at this point. However, because the impacts of changed climate and weather patterns could be so great, it is important to think through contingency planning. One strategy may be to significantly reduce greenhouse gas emissions from human activity. If this strategy becomes policy, a premium will be placed on energy conversion and end-use technologies that reduce these emissions. The committee has not extensively analyzed greenhouse gas emissions for different fuels and feedstocks. However, to give some perspective on the relative contributions of different alternatives, some data from the literature are presented.

The raw material production, manufacture, transportation, and combustion of fuels produces greenhouse gases—carbon dioxide (CO_2), methane (CH_4), nitrous oxide (N_2O), and ozone (O_3)—in amounts depending on the fuel and technologies used. Calculations have recently been made of the production of greenhouse gases from different fuels and vehicles, considering emissions from the entire fuel cycle (DeLuchi et al., 1989; DeLuchi, 1989). The results are expressed in CO_2 equivalents by converting the mass emissions of those gases other than CO_2 into the mass amount of CO_2 that would have the same temperature effect. *Same temperature effect* is defined in degree-years over a given period, where 1 degree-year is an increased surface temperature of 1°C for 1 year. Differences in vehicle efficiency are also factored in. Results are expressed as CO_2-equivalent emissions compared to those vehicles using gasoline and diesel derived from petroleum (Table 5-1).

Transportation fuels such as gasoline, diesel, alcohols, and CNG can be manufactured from plant matter such as wood. If no nonrenewable carbon fuel is used in growing, harvesting, or manufacturing these fuels, at steady state there should be no net contribution of the greenhouse gas CO_2. However, this is by no means the situation for energy-intensive crops such as corn that require major energy inputs for planting, fertilizing, harvesting, and drying.

CNG vehicles using fossil methane have a lower production of greenhouse gases than gasoline-powered vehicles, although this result depends on assumptions about the relative contributions of CH_4 and CO_2 to global warming and on CH_4 emissions from production and use (DeLuchi, 1989; Ember et al., 1986). In this particular analysis, CNG is 19 percent better

TABLE 5-1 Approximate Greenhouse Gas Emission per Mile Relative to Petroleum-Powered Internal Combustion Engines

Fuel and Feedstock	Percent Change
Current Technology	
CNG, gasoline, diesel, or methanol from biomass[a]	−100
Gasoline and diesel from crude oil[b]	0
CNG from natural gas[c]	−19
Methanol from natural gas[c]	−3
Gasoline from oil shale[d]	27 to 80
Methanol from coal (baseline)[c]	98
Potential Advanced Technology	
Gasoline from coal or shale using nonfossil sources for process heat and hydrogen[e]	0

[a] Percent change is for CO_2 only. This is true only for biomass processes that do not use fossil fuel, that do not displace land from forest that would otherwise sequester carbon in its biomass, and that are grown every year so that carbon dioxide from fuel use is taken up by the crops.

[b] Should be increased by 25 to 33 percent for thermally enhanced oil recovery.

[c] The analysis considered emissions of CH_4, N_2O, and CO_2 from the production and transportation of the primary resource (coal, natural gas, or crude oil); conversion of the primary resource to transportation energy (e.g., natural gas to methanol); distribution of the fuel to retail outlets; and combustion of the fuel in engines, except as noted. N_2O emissions from vehicle engines were not included. Emissions of ozone precursors, chlorofluorocarbons (CFCs) from air-conditioning systems, and water (H_2O) were not considered (available data and models do not allow estimation of the greenhouse effect of emissions of ozone precursors; CFC emissions are independent of fuel use; and H_2O emissions from fossil fuel use worldwide are a negligible percentage of global evaporation). The composite greenhouse gas is actual mass emissions of CO_2 plus CH_4 and N_2O emissions converted to mass amount of CO_2 emissions with the same temperature effect.

[d] Considers only CO_2.

[e] Nonfossil sources could be biomass, nuclear, or solar energy devices.

SOURCE: Adapted from DeLuchi et al. (1988a) and DeLuchi (1989).

than petroleum-based gasoline; however, the methane equivalency factor is uncertain, and varying it from 5 to 30 (instead of using 12) would cause the CNG impact to be –4 to –25 percent (DeLuchi et al., 1988).

Because of coal's lower hydrogen-to-carbon (H/C) ratio, using coal as a feedstock with current technologies for transportation fuels would increase greenhouse gas emissions significantly. However, if nonfossil sources of energy were used for hydrogen production and process heat for the conversion processes, the net effect of coal-based fuels would be about the same as for fuels from petroleum.

CO_2 emissions from shale conversion would vary widely depending on the process technology. Western oil shale kerogen has a higher hydrogen and lower oxygen content than coal, resulting in less CO_2 emissions from hydrogen and heat generation. However, decomposition of shale carbonates can release CO_2. Calculations indicate that carbonate decomposition of shale rock would be held below 10 percent from a hot solids retorting process, releasing about 24 gC/megajoule as CO_2, a 27 percent increase over 19.2 gC/megajoule for burning petroleum. Modified in situ (MIS) processes decompose a large fraction of the carbonate rock and 100 percent decomposition releases about 35 gC(as CO_2)/megajoule, about 80 percent more than petroleum. Thermally enhanced oil recovery would produce more CO_2 than conventional petroleum recovery because it typically takes about 25 to 33 percent of the oil produced for thermal heating. This increase would not occur if nonfossil sources of heat were used.

Schulman and Biasca (1989) calculated CO_2 emissions for different fuels in terms of pounds of CO_2/million Btu. Their results also show significant increases of using coal in comparison to petroleum or natural gas but do not include the other greenhouse gases.

There are significant uncertainties in all these calculations. However, in general, feedstocks of lower H/C ratios generate more greenhouse gas emissions. For conversion processes using coal for process heat and hydrogen production, coal-based fuels look the least attractive for limiting greenhouse gases. However, successful R&D on conversion processes and use of nonfossil energy for process heat and hydrogen production can reduce the impact to the equivalent of that from petroleum-based fuels and of methanol from natural gas. In any long-term evaluation of greenhouse gas strategies, consideration of the contribution of the entire transportation sector to the global effect and the various trade-offs involved in fuel manufacturing and in switching to alternative fuels, vehicles, or transportation systems is needed.

Conclusion

Because manufacture of transportation fuels from coal and oil shale resources produces more CO_2 than processes based on oil, natural gas, or nonenergy-intensive biomass, a special effort should be made to identify and pursue opportunities for reduction in CO_2 emissions from these sources.

Biomass could supply the hydrogen or heat for fossil fuel conversion processes. Since biomass supply will probably limit its use, system studies of the optimum use of biomass for reduction of CO_2 emissions from fossil fuel conversion are recommended. In the longer term other nonfossil energy sources for heat and hydrogen production should also be investigated.

6
Major Conclusions and Recommendations for R&D on Liquid Transportation Fuels

OVERVIEW

World oil resources are large enough that low-cost production of oil is expected well into the 21st century, although cartel action will likely keep international oil prices substantially above cartel production costs. While the United States has plentiful fossil fuel resources, production costs for transportation fuels derived from most of these resources are currently greater than those from most imported petroleum. The level of oil prices of recent years, combined with the expectation of continued price volatility, has substantially decreased private investment in exploration, development, and research on domestic resources. In time, however, imported oil prices may increase to the point where a large portion of U.S. domestic resources are again attractive, especially if the cost from domestic resources can be lowered.

A continued decline in private investment in domestic oil and gas production is expected over the near term, however. Without government assistance this decline will result in continually decreasing domestic production. Government assistance can take two forms: (1) improved financial incentives for investment in domestic production and (2) support of research, development, and demonstration of technologies for lower-cost production from domestic resources.

The first form of assistance would be needed if the United States chose to slow the near-term decline in oil and gas production and to stimulate the use of advanced techniques for increasing resource recovery. Over the longer term, support of continuous R&D in advanced oil and gas recovery and in pioneering coal liquids and shale oil developments would accelerate the use of these resources.

This study focuses on the second of these approaches. Emphasis is placed on production of carbon-containing liquid transportation fuels and the use of fossil combustion heat to drive the processes. Under Scenario V (controls on greenhouse gas emissions), presented in Chapter 1, energy alternatives other than fossil fuels would need to be considered. These are being addressed in a concurrent study by the National Academy of Sciences, Committee on Alternative Energy R&D Strategies. Reduction of carbon dioxide emissions for processes considered in the present report could be accomplished by using nonfossil sources of energy for process heat and hydrogen production. These are discussed briefly later, but the committee was not able, because of time constraints, to investigate these processes in technical detail.

Federal R&D at the U.S. Department of Energy (DOE) is an important factor in advancing technology to decrease the costs and environmental impacts of producing liquid transportation fuels from domestic resources. Several issues must be considered in establishing the nature and size of a DOE R&D program for producing such fuels:

- expected timing of commercial application;
- potential size of the application;
- potential for cost reduction, improvements in reliability; and diminished environmental impacts; and
- the need for DOE participation.

R&D Issues

Timing of Commercial Applications

The timing of commercial application of new technology depends critically on production costs and environmental impacts. These costs depend on the technology but are also strongly influenced by environmental considerations and by state and federal taxes and tax credits. The scenarios for the future presented in Chapter 1 were developed to provide a framework for the committee's recommendations. The scenarios cover a range of possibilities, whose relative probabilities are inevitably matters of judgment.

For research planning the committee judged the most probable economic scenario to be that future oil prices would be between $20 and $30/barrel (in 1988 dollars) within 20 years from now (Scenario II). However, the likelihood that prices would either remain under $20/barrel or exceed $30/barrel appears high enough to necessitate program recommendations that are reasonable given any of the three scenarios presented in Chapter 1.

Potential Size of the Applications

Potential size of the applications depends on the size and geographical distribution of the resource. A geographically dispersed resource offers more widespread commercial and employment opportunities and is less vulnerable to local disruption, regulations, and restrictions.

Potential for Cost Reduction

The potential for cost reduction is generally least for mature and technically advanced operations. However, for very large scale activities, such as oil and gas production, even small percentage improvements can justify extensive research.

Need for DOE Participation

U.S. R&D in transportation fuels production is the sum of industry-supported and government-supported activities. The role of DOE is to help ensure that the major national needs for technology are met and that potential benefits of domestic production, not the subject of R&D by private firms, are pursued where justified by the above criteria and can be realized. Where there is substantial and continuing industrial involvement, the role of DOE is generally to support long-range and relevant basic research and in some cases to participate in large demonstration programs, such as the Clean Coal Program. In areas where commercial projects are far in the future but where continued technological advances are in the national interest, it is logical for DOE to take a lead role.

RESOURCES

Petroleum, Heavy Oils, and Tar

Projections of the availability of petroleum, heavy oils, and tar from domestic resources, summarized in Table 6-1, show that, for an oil price of roughly $25/barrel, current production rates could be maintained for some decades. Even lower but stable prices from $20 to $24/barrel would encourage production from resources that are still substantial. A higher oil price ($40 to $50/barrel) would make possible the more extensive development and use of advanced oil recovery techniques.

Scenario I (with future oil prices less than or equal to about $20/barrel) would result in a continued decline in U.S. oil production, while in Scenario II (prices reach $30/barrel within 20 years) or Scenario III (prices reach

TABLE 6-1 Estimated Remaining Economically Producible Crude Oil Resources[a]

	Current Technology		Advanced Technology	
	$24-$25[b]	$40-$50[b]	$24-$25[b]	$40-$50[b]
Billion barrels oil	75-76	95-140	105-129	140-247
Ratio of resource base to annual production	25	32-47	35-43	47-82

[a]See also Chapter 2.
[b]Oil price ($/barrel).

$40/barrel within 20 years), U.S. oil production decline could be reduced for at least the 20-year period of the scenario.

Scenario IV (imposition of more stringent environmental controls) seems quite probable. In general, greater environmental controls will increase the costs of exploration and production and will delay the application of advanced oil recovery techniques. Closing frontier areas for exploration and production also reduces the amount of oil available at a given price and shortens the time over which domestic oil could supply a major fraction of U.S. transportation fuels. These trends would increase the need for imports. Energy efficiency improvements can be very important and can help reduce imports. Scenario V (greater greenhouse gas controls) would tend to mitigate against thermal enhanced oil recovery and CO_2 enhanced oil recovery using fossil CO_2.

For Scenario VI (no government encouragement of domestic oil production), U.S. oil production decline will continue for oil prices below $20/barrel. Even under the price increases of Scenario II and Scenario III, the stabilization of production would require years. Thus, if the U.S. government wanted to retard domestic oil production declines, some form of government encouragement would be required.

Not only is U.S. petroleum production declining, but industry emphasis is changing. The major oil companies are increasingly investing abroad where costs are lower, the potential for successful large oil fields is higher, and where developing countries are offering special incentives to encourage development of their petroleum resources. In addition, the number and financial health of small independent companies and individuals has decreased.

While the traditional form of industry encouragement is through tax incentives, improved technology and its transfer through cooperative efforts will be of increasing importance, especially for the independent operators who, in general, do not have significant R&D programs. To the extent that licensing of technology and use of expert consultants do not facilitate technology transfer to the independent sector, significant advice may be necessary to develop and make available advanced technology to this segment of the domestic oil-producing industry.

Oil cannot be produced to exhaustion at a constant rate but generally declines slowly over time. Even if constant fuel consumption could be maintained, it seems reasonable to expect that significant supplemental sources (either domestic or imported) of transportation fuels will be needed 20 to 30 years from now. At this time it is expected that R&D on fuels from coal and oil shale would reduce the costs to the level where they could compete with petroleum-based fuels.

Natural Gas and Synthesis Gas

Economically producible resources of natural gas for two price levels and different levels of technologies are summarized in Table 6-2. An expansion in the resource base of more than 100 percent is projected at the high price, given use of advanced technology. Significant amounts of gas could therefore be made available as an alternative source of transportation fuels. There are several approaches to exploit this resource:

- use compressed gas directly for transportation fuel;
- displace fuel oil from power generation and industrial fuel use, making it available for conversion to transportation fuels;
- manufacture hydrogen and carbon monoxide for production of transporation fuels; and
- possibly use advanced, low-cost processes for direct conversion to liquid transportation fuels.

Compressed natural gas vehicles, while not expected to be a significant part of the market because of short vehicle range and onboard storage constraints, have recently attracted much interest as a relatively low polluting alternative for urban fleet use.

TABLE 6-2 Estimated Remaining Economically Producible Natural Gas Resources

	Current Technology		Advanced Technology	
	$3[a]	$5[a]	$3[a]	$5[a]
Tcf Gas (Bbbl oil equivalent)	595 (107)	770 (140)	880 (160)	1,420 (256)
Ratio of Resource Base to Current Production	33	43	50	80

[a]Wellhead gas price ($/Mcf).

A rise in oil price to the range of $25 to $40/barrel would make conversion of heavy fuel oil and heavy oil to transportation fuels more economically attractive. This use is expected to grow. Tar sands bitumen could also be upgraded. Natural gas could be used as the hydrogen source for hydroconversion (which increases liquid yields) of these heavy fuels. Natural gas consumption would also increase from the replacement of the heavy fuel oil that might otherwise be used in power generation and industrial boilers and heaters.

Coal liquefaction and methanol and Fischer-Tropsch (F-T) liquid synthesis from coal are also potentially very large consumers of hydrogen or synthesis gas. For example, the production of the equivalent of 1 billion bbl/year (2.74 MMbbl/day) of crude oil (30 percent of current production) would require about 40 percent of the gas now produced. Such an increase could come from domestic resources, but it would require greatly accelerated exploration and production and would increase the cost of natural gas.

Methane would likely be used for hydrogen in the initial stages of fuels manufacture from coal and shale because gas price increases will probably lag oil price increases, and gas prices may be initially lower than those of the base case used in the economic studies. For the longer term, however, coal gasification may be more economical than use of natural gas to supply hydrogen for coal and shale liquefaction.

Estimated costs for alternative conversion processes to make transportation fuels are shown in Table 6-3. Table 6-3 also illustrates the extreme sensitivity of methanol costs to natural gas costs. In the process of producing methanol, coal or natural gas is first converted to synthesis gas. The conversion of coal to syngas is a major cost; high natural gas prices also make natural gas conversion to syngas expensive. Methanol synthesis consumes more synthesis gas than coal liquefaction and tends to be more expensive for equal synthesis gas costs. The natural gas price of $4.89/Mcf, corresponds to the historical domestic relationship between gas and fuel oil prices and to a price where coal gasification is expected to be competitive as a methanol source. For the lower price of $3.00/Mcf, gasoline from heavy oils is competitive. At still lower prices, corresponding to low value remote gas, imported methanol would be less expensive than methanol produced domestically from higher priced gas indicating that, unless assisted by legislative action based on environmental and energy security consideration, fuel methanol will be imported.

For the longer term, when natural gas will be expensive, advances in the manufacture of synthesis gas from coal could reduce methanol costs, and a DOE program in syngas manufacture from coal for use with coal liquefaction could then make a major contribution if it becomes desirable to produce methanol domestically.

TABLE 6-3 Equivalent Crude Oil Cost of Alternative Fuels (in 1988 dollars/barrel, at 10 percent discounted cash flow)

Process	Current Cost Estimates[a]	Cost Targets for Improved Technology
Heavy Oil Conversion	25	—
Coal Liquefaction	38	30
Coal/MTG	62	—
Western Shale Oil	43	30
Methanol		
Coal gasification[b]	53	—
Natural gas at[c]		
$4.89/Mcf	45	—
$3.00/Mcf	37	—
$1.00/Mcf	24	

[a] The processes on which these numbers are based are in various stages of R&D (see Chapter 3).
[b] See Table D-4a. New estimated reduced capital and operating expenses for entrained-flow coal gasification could lead to coal-to-methanol costs of about $40/barrel.
[c] See Table D-7.

Coal and Oil Shale Conversion

Direct coal liquefaction is shown in Table 6-3 to have a lower estimated cost than oil from western shale. In the past the estimated costs for conversion of oil shale were somewhat lower than for coal liquefaction. This change in relative costs reflects the progress from steady DOE R&D on coal liquefaction in recent years. The committee believes that vigorous R&D and optimizing these processes has the potential to bring the cost of both down to about $30/barrel or lower. A substantial effort is required to accomplish this reduction, and, with the reduced industry effort, government encouragement through DOE participation and leadership is essential. The price assumption for Scenario II ($30/barrel) allows approximately 20 years to demonstrate coal or oil shale processes that can compete with $30/barrel oil. This scenario is consistent with an R&D program organized around a $30/barrel goal and a large pilot plant and demonstration programs arranged when pathways to reaching this goal have been established.

The coal liquefaction program has a good start in this direction; however, since achieving this cost reduction is aided by improvement of and integration with the coal gasification process, some strengthening of coal gasification research is indicated if the potential for sizeable cost reductions can be realized.

The shale program has recently received much less attention than coal liquefaction. While the size of the shale resource is comparable to that of coal, the active industry and geographical dispersions of coal resources are consistent with giving a somewhat higher priority to coal. An increase in the shale oil program, however, is needed to bring the two programs into better balance.

ENVIRONMENTAL CONSIDERATIONS

The manufacture and use of transportation fuels raise many environmental issues. For fuel production, air and water pollution can generally be controlled to meet emission standards with available technology, although lower-cost technologies are needed. Special problems include the prevention of significant deterioration of air quality in some regions and the recovery of solvent in tar sands extraction. Issues related to land use and visual impacts are beyond the scope of the current study and must be addressed in the political and regulatory arena. R&D efforts must change with social priorities.

The contribution of gasoline vehicle emissions to urban air pollution has generated increased interest in alternative-fueled vehicles using, for example, natural gas or methanol. An alternative may be reformulating gasoline to facilitate redesign of improved vehicle emission control systems. Future vehicle emissions constraints may well affect fuel composition and therefore the choice of conversion processes and related research programs.

Although environmental restrictions may influence automotive fuel composition, the economic, environmental, and health effects of fuel components (paraffins, aromatics, methanol, formaldehyde, and other oxygenates) and optimal control technologies and engine designs are far from well established. The DOE should participate in quantifying these effects and variables to help ensure that production technologies for liquid transportation fuels from domestic resources are properly developed to meet future regulations on vehicle emissions. This area requires more detailed study.

The greenhouse effect is of increasing concern, and the production and use of transportation fuels could be an increasing source of CO_2 and other greenhouse gases. Table 5-1 shows estimates of relative greenhouse gas emissions for the manufacture and use of transportation fuels from several sources.

Because of coal's low hydrogen content and impurities, the manufacture and use of liquid fuels from coal produce almost twice the CO_2 as use of gasoline from petroleum (see Table 5-1). Manufacture and use of liquid fuels such as methanol or F-T gasoline from methane, however, produce an amount of CO_2 approximately equal to that from petroleum-based gasoline. Gasoline from oil shale produces less CO_2, the amount depending on the amount of decomposition of carbonates during retorting. CO_2 emissions can be reduced in all cases by increasing end-use efficiency and by reducing process heat requirements.

The heat necessary to drive processes is conventionally derived from combustion of fossil fuel, with liberation of CO_2. In addition, hydrogen needed for processing is derived from water, where oxygen is eliminated by CO_2 generation. This is also a major CO_2 source beyond the CO_2 generated by fuel end-use. Nuclear or solar energy and biomass are alternative sources of heat and hydrogen. Water splitting by heat, electrolysis, or photolysis using noncombustion sources of energy is substantially more expensive than use of carbon as an oxygen acceptor (NRC, 1979). However, a long-range exploratory and basic research program on water splitting is justified.

Use of biomass to supply heat and hydrogen to fossil fuel processes (if use of fossil fuels in biomass production and processing is minimized) can eliminate or reduce these sources of CO_2. Comparison of the use of biomass-generated methanol via synthesis gas to the conversion of this synthesis gas to hydrogen and its use for coal liquefaction indicates that, for a limited supply of biomass, a greater reduction of fossil carbon-generated CO_2 is obtained by combining biomass gasification with coal liquefaction. System studies research relevant to this combination are recommended.

MAJOR CONCLUSIONS AND RECOMMENDATIONS

A federally funded R&D program on liquid transportation fuels can provide future options for domestic uncertainities in oil prices and investment decisions by the private sector. The current funding for liquid fuels R&D is only about 29 percent of the total fossil energy budget (see Table 6-4). A diverse approach to different resources and technologies expands these options in recognition that there may be failure of some technologies and resources may fail to meet expectations.

The DOE program should contain a continuing effort by unbiased and capable groups to evaluate the economic and commercial potential of the technologies in the program. The most promising technologies should be moved forward from the research laboratory to field test units and eventually to larger facilities for demonstration on small commercial equipment. The government-sponsored program should include industrial participation at all phases, particularly in development and demonstration to facilitate technology transfer and ensure that the latest practical industrial concepts are incorporated into the program. A properly balanced program should

TABLE 6-4 DOE's Office of Fossil Energy R&D Program Budget (current dollars in millions)

	FY 1988 Appropriations	FY 1989 Appropriations	FY 1990 Request	FY 1990 House	FY 1990 Senate Panel
Coal Budget					
Control technology and coal preparation	$43.62	$48.93	$32.26	$60.10	$53.13
Advanced technology R&D	24.94	25.56	25.54	26.18	29.32
Coal liquefaction	27.13	32.39	9.66	37.68	33.26
Combustion systems	25.17	26.70	15.77	35.27	30.17
Fuel cells	34.20	27.53	6.50	38.40	29.80
Heat engines	17.95	22.83	8.92	20.02	21.22
Underground gasification	2.78	1.37	0.43	0.43	0.83
Magnetohydrodynamics	35.00	37.00	0	42.90	37.00
Surface gasification	22.99	21.56	8.74	19.64	29.88
Total coal	$233.78	$243.87	$107.82	$280.62	$264.61
Petroleum Budget					
Enhanced recovery	$16.54	$23.58	$18.24	$27.59	$28.46
Advanced process technology	3.43	4.20	4.62	3.60	3.60
Oil shale	9.50	10.53	1.68	8.18	10.88
Total oil	$29.47	$38.31	$24.54	$39.37	$42.94
Gas Budget					
Unconventional gas	$10.53	$11.38	$4.07	$13.17	$15.82
Cooperative R&D Ventures	$0	$0	$0	$4.80	$4.80
Total gas	$10.53	$11.38	$4.07	$17.97	$20.62
Miscellaneous[a]	$53.22	$88.03	$26.15	$84.72	$81.17
Total fossil R&D	$327.00	$381.59	$162.58	$422.68	$409.34

[a] Includes plant and capital equipment, program direction, environmental restoration, fuels conversion, and past year's offsets. Numbers may not add due to rounding.

SOURCE: July 31, 1989, Clean-Coal/Synfuels Letter.

achieve a key objective of providing an understanding of how U.S. resources can best be used to produce transportation fuels.

Since industry participation is essential to an effective program, DOE must provide the appropriate leadership to achieve such participation. The DOE can encourage industrial participation by proper structuring of the program. In addition, industrial participation will be more readily achieved if the DOE R&D program is viewed as a key element of a national energy policy.

A steady program is essential for success in a long-range R&D program. To ensure such a program there must be a long-term funding commitment, and the elements of the program should be primarily decided by DOE technical and administrative professionals based on technical and economic merit.

Furthermore, a well-balanced program should include demonstration of state-of-the-art technology on small-scale commercial equipment as well as continued search for new technology that may eventually make the demonstrated technology obsolete. In this way the program will continually be updating information on the best way to use domestic resources for transportation fuels. Technology that is selected for demonstration must meet strict economic and environmental criteria. Thus, the portion of the program devoted to demonstration is determined largely by opportunities generated by the research program and further developed in pilot facilities. The program should develop specific objectives within the next 5 years regarding demonstration of the best technologies.

The recommended directions for the DOE program for the next 5 years are listed below in three funding categories. The listing within each category is not in priority order. All areas listed are of potential importance and there should be continuing related programs of fundamental and exploratory research. The need for the more costly process R&D depends on the need for DOE participation during the next 5 years and varies considerably.

Under Scenario II the premise that oil prices will reach $30/barrel within 10 to 20 years conforms with a target of $30/barrel for coal and oil shale through pilot projects and studies over the next 5 years. Under Scenario I the pace of the program could be slowed, whereas Scenario III would call for a more rapid pace. Increased emphasis on curtailing greenhouse gas emissions would result in techniques for reducing such gases, whereas greater environmental constraints would lead to emphasis on environmental research. The recommended areas of R&D as proposed are diverse and provide options in the face of the uncertainty these scenarios encompass.

Major Funding Areas

With regard to use of domestic resources, the high funding areas are oil and gas, coal, and oil shale. These represent large domestic resources with oil R&D also providing a means to significant U.S. production over a period of time when coal and oil shale technologies can be further developed. The committee has not made a detailed analysis of required federal funding for R&D activity for these resources. However, they are all of major importance, and this should be reflected in their relative funding levels.

There is less need for DOE funding of R&D on conventional gas production since there is much private sector activity but DOE should continue its work on unconventional gas recovery. Significant funding and attention is also recommended for R&D related to fuel composition and its environmental and end-use consequences.

1. *Participation in R&D and Technology Transfer for Oil and Gas Production.* In recent years the DOE research program in oil and natural gas has been substantially reduced. Industry activity in R&D for domestic oil is also declining. Important opportunities for both cost reduction and improved resource utilization exist, and DOE participation should be in balance with other energy research areas. The program should focus on those parts of the resource base whose exploitation depends on more comprehensive understanding of geologically complex reservoirs and on technologies yet to be fully developed. The program should be pursued in coordination with industry (both independent oil producers and major oil companies), preferably with direct industry participation. Finally, an effective program of information and technology dissemination is needed.

2. *Production from Coal and Western Oil Shales.* Coal and western oil shales both represent very large resources compared to domestic petroleum and natural gas. Estimated costs with current technology require oil prices greater than $36 to $43/barrel, but recent advances suggest that their costs may be reduced to the equivalent crude oil price of around $30/barrel or less. Because the cost of producing domestic oil may rise to this level in the next several decades and this is also the time frame required to bring new technology to commercial status, DOE should establish the goal of reducing the cost of these alternatives to below $30/barrel. The DOE should also take the lead in establishing a demonstration program when pilot plant and engineering studies indicate that this goal can be achieved. Important components of such a program are the following:

- a vigorous basic and exploratory research program;
- a pilot plant program capable of supplying the information needed for commercial-scale designs;
- continuing systems studies aimed at optimization;
- a new thrust aimed at integration of hydrogen production from both biomass and coal; and
- a high level of industrial involvement.

Over the next 5 years, exploratory research on coal should stress new catalysts and processes based on fundamental coal science understanding. The opportunity to reduce costs by integrating hydrogen manufacture should be explored. The program should be guided partly by economic and technical evaluations by engineering firms, petroleum industry operating companies, and qualified consultants. The program should have a 5-year objective to reduce the cost of direct liquefaction to $30/barrel or less. If this objective is achieved, preparations for a larger pilot plant (500 to 1000 bbl/day) would begin.

In the judgment of the committee, the current shale oil program is too small compared to the coal liquefaction program and should be increased. Over the next 5 years a field pilot facility with a capacity of about 100 bbl/day should be built to further develop surface retorting technologies. These technologies must clearly have the potential for meeting anticipated environmental requirements and for production costs of $30/barrel or less.

Because manufacture of transportation fuels from both of these resources produces more CO_2 by-product than processes based on oil, gas, or biomass, a special effort should be made to identify and pursue opportunities for reduction in emissions of this greenhouse gas from these resources. Study of nonfossil fuel sources of heat and hydrogen should be included.

3. *Environmental and End-Use Considerations.* There are a number of uncertainties about the health, safety, and air quality implications of alternative fuels use. With other federal agencies, such as the U.S. Environmental Protection Agency and the National Institutes of Health, DOE should continue R&D to develop a better data base on these potential impacts. In particular, health effects and also different fuel-engine- emission controls combinations should be investigated to identify the safest and most cost-effective combinations and to provide guidance on fuel composition effects for use in the DOE R&D programs. This will help ensure that future regulations are balanced and on a firm technical basis and that the technologies for liquid transportation fuels production are properly developed to meet these regulations.

Medium Funding Areas

4. *Coal-Oil Coprocessing.* Coprocessing of heavy oils or residuum with coal may permit the introduction of coal as a refinery feedstock. It is expected to have rather limited application unless important synergism between oil and coal occurs. Funding of basic bench-scale research should be continued over the next 5 years to define the extent of synergy between coal and oil for coprocessing coal-residuum combinations, followed by a thorough economic analysis of its impact. If favorable, the results should be confirmed in the Wilsonville test facility to define optimal processing conditions.

5. *Tar Sands.* The domestic tar sands resource is small relative to those of coal and oil shale. However, it is significant relative to proven domestic crude oil reserves, and much of it is owned by the federal government. Liquid fuels can potentially be produced from some U.S. tar sands at about $25 to $30/barrel equivalent crude oil price with a hydrocarbon solvent extraction process. Furthermore, there is little industry activity in this area. Therefore, a modest DOE R&D program on tar sands is appropriate if there are sufficient leads toward cost reduction or if costs are low enough to justify development and demonstration of the best technology.

Over the next 5 years all potential processes and mining techniques applicable to U.S. tar sands should be evaluated both technically and economically. The DOE should sponsor preliminary evaluations by engineering firms, petroleum operating companies, and qualified consultants. The best process should be selected for further development and demonstration in a field pilot plant with a capacity of 50 to 100 bbl/day. Based on Canadian experience, this size should be suitable for scale-up to a commercial plant. A field pilot operation is justified only if the technology is judged to be sound, all environmental requirements are projected to be met, and costs are sufficiently low (probably about $25/barrel) to attract industry participation.

6. *Petroleum-Residuum, Heavy Oil, and Tar Conversion Processes.* Conversion processes for petroleum residuum, heavy oils, and tar have been under intensive development in both domestic and foreign petroleum industries. Increasing crude oil prices will tend to favor hydroconversion processes over carbon rejection processes because of the higher liquid product yield from hydroconversion. This continuing industrial process development should be supplemented by basic research on the molecular structures of metals, sulfur, and nitrogen-binding sites and coke precursor species in heavy oil feeds and upgraded products. Results of this research would help the private sector improve existing carbon removal and hydrogen addition processes. The DOE should involve the private sector in the design of this research program to ensure good technology transfer. This R&D area is considered medium priority because there is considerable activity in the private sector. Because of industrial efforts, DOE work on catalyst and process development is not recommended at this time.

7. *Biomass Utilization.* Use of some biomass resources for the production of liquid transportation fuels is one pathway that can result in less net release of greenhouse gases. Biomass supply constraints and costs will probably require continued use of fossil fuel resources. Use of biomass to produce liquid fuels directly is of continuing interest; however, by integration of processing of biomass and fossil resources (e.g., by generating process hydrogen from biomass instead of coal), a greater reduction in CO_2 from the combined processes may be achievable. There is little industry

activity in this area. Hence, it is recommended that research and systems studies be conducted on the optimum integration of biomass with fossil fuel conversion processes as well as for stand-alone biomass conversion.

8. *Coal Pyrolysis.* The current DOE program is aimed at production of pyrolysis liquids and metallurgical coke and does not have a high priority for liquid transportation fuels.

There is little privately funded R&D in this area. The chemistry and mechanisms of pyrolysis are not well understood, and therefore DOE should place medium priority on a program of basic pyrolysis research, including research in catalytic hydropyrolysis. Systems studies should be carried out over the next 5 years to evaluate integrating pyrolysis with direct coal liquefaction as well as with gasification or combustion.

Modest Funding Areas

9. *Processes for Producing Methanol, Methanol-derived Fuels, or Fisher-Tropsch (F-T) Liquids from Synthesis Gas.* Industry is vigorously studying the production of methanol and F-T Liquids. While they may find application in the United States, production is expected primarily outside the United States where low-cost natural gas is available. These factors discourage DOE work in this area beyond fundamental and exploratory research.

10. *Direct Methane Conversion.* Direct methane conversion to liquid hydrocarbons or methanol is being studied at the bench scale by various companies, government agencies, and universities. These processes theoretically have the potential for being more energy efficient and less expensive than indirect conversion since they bypass the formation of syngas, an energy-intensive and expensive step. However, potentially significant cost reductions have not yet been achieved.

Even if direct conversion of natural gas to liquid fuels becomes economically viable, the sources would be predominately low-cost natural gas in foreign locations. U.S. government-sponsored research on direct methane conversion technology should be limited to continuing fundamental and exploratory research.

11. *Eastern Oil Shale.* Although widespread, most eastern oil shale is low grade, occurs in thin seams, and has a high stripping ratio for mining. Its processing is also inherently more expensive than that of western shale because of its low grade and low hydrogen and high sulfur content. These disadvantages are expected to outweigh the infrastructure advantages of the eastern location. This resource will be economical only after exploitation of coal or western oil shale. No development is recommended at this time.

Appendixes

A

Statement of Task

1. The committee will review the most recent and/or most complete technology assessments and cost estimates for the production of liquid transportation fuels from coal, oil shale, natural gas, petroleum from enhanced oil recovery, and tar sands. Consultants will be used to help the committee in this process.

2. The committee will conduct workshops as needed to evaluate the work of the consultants and to assess the cost estimates and the various R&D opportunities that exist to reduce the costs of the conversion technologies. Experts will be invited to these workshops as appropriate.

3. The committee will develop a summary assessment of each technology with regard to (1) the stage of technology development relative to that needed for commercial deployment, (2) the credibility of the cost estimates based on the nature of the estimate and the stage of technology development, (3) the "fit" between the characteristics of the products produced and marketable liquid fuel requirements, and (4) the potential for improved technology and cost reduction based on continuing R&D.

4. The committee will assess the potential commercial impact of each technology based on the above findings.

5. The committee will identify the major technical and other barriers to the commercial deployment of each technology, including supply, distribution, environmental, delivery, and end-use considerations.

6. The committee will estimate the time frame in which each technology would be expected to be commercially deployed. These judgments should be based on the most recent petroleum and gas supply, demand, and price projections provided by the Energy Information Administration in its yearly Annual Energy Outlook volume.

7. The committee will prepare a 5-year overall strategic and program plan for a "broad research program" for liquid transportation fuels based on plentiful supplies of domestic resources. This plan is *not* intended to be an implementation plan that would include specific R&D tasks, milestones, and associated funding.

B

Committee Meetings and Activities

1. **Committee Meeting, March 14-15, 1989, Washington, D.C.**
   Presentations made to the committee by Jay Braitsch, U.S. Department of Energy.

2. **Committee Meeting, May 8-9, 1989, Washington, D.C.**
   The following presentations were made to the committee:
   (a) Overall Structure and Organization of DOE's Fossil Energy Program, Jay Braitsch, U.S. Department of Energy (HQ).
   (b) Overview of DOE's Coal Technology R&D Program, Gary Voelker, U.S. Department of Energy (HQ).
   (c) Direct Coal Liquefaction (Including Coprocessing and Advanced Technologies), Gil McGurl, Pittsburgh Energy Technology Center (PETC).
   (d) Indirect Coal Liquefaction, Gil McGurl, PETC, and Paul Scott, DOE (HQ).
   (e) Coal Gasification, Larry Rath, Morgantown Energy Technology Center (METC).
   (f) Coal-Fired Diesels and Turbines for Transportation, Larry Rath, METC.
   (g) Overview of DOE's Oil, Gas, and Shale R&D Program, Marvin Singer, DOE (HQ).
   (h) Enhanced Oil Recovery, George Stosur, DOE (HQ).
   (i) Tar Sands, George Stosur, DOE (HQ).
   (j) Oil Shale, Jerry Ramsey, DOE (HQ).
   (k) Unconventional Gas Recovery, James White, DOE (HQ).
   (l) Advanced Extraction and Process Technology, David Beecy, DOE (HQ).

(m) Underground Coal Gasification for Transportation Fuels, Art Hartstein, DOE (HQ).
(n) Bernard Schulman, SFA Pacific, Inc., consultant to the committee, summarized progress on his work.
(o) Velo Kuuskraa, ICF Resources, consultant to the committee, summarized progress on his work.
(p) Possible Application of Nuclear Power for Process Heat, Paul Kasten, Committee Member.

3. **Committee Meeting and Workshop, June 8-10, 1989, Washington, D.C.**
The following presentations were made to the committee:

(a) Scenarios Postulated by Committee, Robert Hirsch, Committee Member.
(b) Economically Recoverable Reserves and R&D Opportunities for Petroleum, Heavy Petroleum, Tar Sands, and Natural Gas, Velo Kuuskraa, ICF Resources.
(c) Costs and R&D Opportunities for Converting Feedstocks into Transportation Fuels, Bernard Schulman, SFA Pacific, Inc.
(d) Distribution and Marketing of Gasoline Versus New Fuels, Ted Wagner, Consultant.
(e) Engine Issues Related to Gasoline Versus New Fuels, Charles Amann, General Motors Research Laboratories.
(f) Engine Issues Related to Diesel Versus New Fuels, John Wall, Cummins Engine Company.
(g) Occidental's Oil Shale Project, Raymond Zahradnik, Occidental Oil.
(h) Unocal Oil Shale Project, Cloyd P. Reeg, Unocal Corporation.
(i) Overview of Coal Liquefaction Technologies and R&D, Harvey Schindler, Science Applications International Corporation.
(j) Direct Coal Liquefaction Research Opportunities, Frank Derbyshire, Sutcliffe Speakman Carbons, Ltd.
(k) Indirect Coal Liquefaction Opportunities, Alex Mills, University of Delaware.
(l) Petroleum Industry Perspective on Liquid Fuels from Coal, Robert Lumpkin, Amoco Corporation.
(m) Utility Industry Perspective on Liquid Fuels from Coal, Seymour Alpert, Committee Member.
(n) Direct and Indirect Cost Reduction Prospects, David Gray, Mitre Corporation.
(o) Overview of Enhanced Oil Recovery Technologies, Lloyd Elkins, Petroleum Consultant.

APPENDIX B   137

(p) Reservoir Characterization: Unswept Mobile Oil, C. R. Hocott, University of Texas.
(q) Thermal Methods for Enhanced Oil Recovery: Heavy Oils, Richard A. Deans, Texaco, Inc.
(r) Miscible Methods for Enhanced Oil Recovery: Light Oils and Immobile Oils, Fred Stalkup, ARCO Oil and Gas Company.
(s) Chemical Methods for Enhanced Oil Recovery: Sweep and Displacement Improvements, G. P. Willhite, University of Kansas.
(t) Petroleum Liquids Availability, William Fisher, Committee Member.
(u) Tar Sands Resources and Recovery Methods, Gene Tampa, Amoco Corporation.
(v) Tar Sands Activities, John Scott, Alberta Oil Sands and Tar Research Authority.
(w) Thermochemical Conversion of Biomass to Methanol, Tom Reed, Colorado School of Mines.
(x) Hydrolysis of Cellulose Biomass to Ethanol, George Tsao, Purdue University.
(y) Liquid and Gaseous Fuel Options for Natural Gas, William Schumacher, SRI International.
(z) Liquid and Gaseous Fuel Options for Natural Gas, Irving Solomon, Gas Research Institute.
(aa) Fuel Distribution: Costs and Other Aspects for Alcohol and Compressed Natural Gas Fuels, Margaret Singh, Argonne National Laboratory.
(bb) Current Knowledge on Environmental Issues Related to Synthetic Fuels Production and Use, Barry Wilson, Battelle Pacific N.W. Labs.
(cc) Air Pollution and Toxicity Impacts of Alternative Transportation Fuels, Allan Lloyd, South Coast Air Quality Management District.
(dd) Martha Gilliland, University of Nebraska, contributed a summary on environmental impacts of fuel production.

4. **Committee Meeting, July 27-28, 1989, Washington, D.C.**

5. **Committee Meeting, August 23-25, 1989, Washington, D.C.**

C

U.S. and World Resources of Hydrocarbons

Estimates of U.S. and other countries' primary resources of hydrocarbons include conventional crude and heavy oils, oil-bearing shales and tar sands, natural gas and natural gas liquids, coal, and biomass. These data represent current estimates of resources available for future exploitation and are expected to change as knowledge of the resources improves.

Resources of fossil fuels are typically characterized in physical units in terms such as proved reserves, indicated additional reserves, inferred reserves, and undiscovered resources, which are defined below. Biomass resources are often expressed in terms of their fuel conversion potential, usually to neat ethanol or methanol. *Proved reserves* are those quantities of resources for which there is reasonable assurance from geologic, engineering, and other data that the resources are recoverable in future years from known locations under existing economic and operating conditions (EIA, 1989a). Thus, proved reserves are economically recoverable by definition (BEG, 1988). *Indicated additional reserves* typically refer to quantities of crude oil judged to be recoverable by enhanced recovery techniques, but where the economic recoverability of the reserves has not been established with sufficient conclusiveness for them to be included under proved reserves (Considine, 1977). The term *inferred reserves,* also referred to as probable resources, is used to denote that part of the identified economic resource base that will be added to proved reserves through extensions, revisions, and new pay zones. Inferred reserves can thus be thought of as resources that reside at the interface of proved reserves and undiscovered resources (BEG, 1988). As the term implies, *undiscovered resources* are just that. Their estimates rely on informed judgment but remain speculative until validated by discovery and development.

For natural gas three additional categories of the resource base have been defined. The first category is known as *extended reserve growth in nonassociated fields onshore*. It represents gas distributed in known fields recoverable through intensive development of heterogeneous reservoirs (i.e., through infill drilling and recompletion of previously bypassed zones). The second category is *gas associated with improved recovery* (and hence reserve growth) *of mobile oil from known reservoirs*. The third category is termed *unconventional gas* (i.e., gas in low-permeability reservoirs, coalbed methane, and shale gas). Subject to economic exploitation, these three categories have the potential to increase significantly the proved reserves of U.S. natural gas (BEG, 1988).

U.S. and world resources of conventional crude oil are shown in Tables C-1 and C-2, respectively, and U.S. reserves of natural gas liquids are shown in Table C-3. World reserves are estimated to be in the range of 800 billion bbl, of which about 60 percent is to be found in the Middle East. Eight countries (Saudi Arabia, Iraq, Kuwait, United Arab Emirates, Iran, all in the Middle East, along with the USSR, Venezuela, and Mexico) account for around 80 percent of the reserves, while the United States accounts for about 6 percent. In terms of resources yet to be discovered, it is estimated that the United States has over 10 percent of the world's total. It is believed that the United States is third in original oil endowment (after Saudi Arabia and the USSR) but fifth in remaining oil because over 60 percent of its estimated total recoverable oil (about 230 million bbl) has already been produced (Riva, 1988).

Current annual U.S. consumption of petroleum products is around 6 billion bbl, of which about 40 percent is imported. World consumption is about 22 billion bbl/year (EIA, 1988, 1989d).

TABLE C-1 U.S. Resources of Crude Oil (billion bbl)

	Proved Reserves	Indicated Additional Reserves	Inferred Reserves	Undiscovered Resources
Lower-48 onshore	16.6	3.1	11.2	20.1
Lower-48 offshore	3.3	0.1	1.6	12.7
Alaska	7.4	0.6	5.5	16.6
Total	27.3	3.8	18.3	49.4

SOURCE: Kuuskraa et al. (1989). In this reference, data on proved reserves and indicated additional reserves are attributed to the U.S. Department of Energy, the Energy Information Administration (Annual Report, 1987); data on inferred reserves and undiscovered resources are attributed to the U.S. Geological Survey's Minerals Management Service (Open-File Report 88-373).

TABLE C-2 World Resources of Crude Oil (billion bbl)

	Proved and Indicated Additional Reserves[a]		Proved Reserves, Cong. Res. Serv.[b]	Inferred Reserves, 1987 Cong. Res. Serv.[b]	Undiscovered Resources, Cong. Res. Serv.[b]	Total Resources, USGS[a]
	Oil and Gas Journal	World Oil				
North America	80.7	84.3	59.8	24.4	89.3	139.1-280.5
Central and South America	65.7	66.9	33.2	19.2	25.2	74.6-135.6
Western Europe	22.4	20.1	19.5	9.4	19.3	45.1-78.1
USSR and Eastern Europe	60.8	61.5	59.0	22.0	77.0	
Africa	55.2	54.2	51.0	11.9	18.5	88.5-158.5
Middle East	564.7	470.6	384.5	47.5	99.0	482.9-619.9
Far East and Oceania	19.4	20.9	36.7	8.1	53.9	41.2-94.2
World Total	868.9	778.5	643.7	142.5	382.2	

[a]EIA (1988, 1989a). The EIA attributes the data on proved and indicated additional reserves to the *Oil and Gas Journal,* Vol. 82, No. 53 (December 28), 1987, and to *World Oil,* Vol. 207, No. 2 (August 1988), and the data on total resources to an assessment by the U.S. Geological Survey as of January 1, 1985. As cited by EIA, total resources include inferred reserves and undiscovered resources and are represented with a probability range of 95 to 5 percent.
[b]Riva (1988). Data from 1987.

TABLE C-3 U.S. Resources of Natural Gas Liquids (billion bbl)

	Proved Reserves	Inferred Reserves
Lower-48 onshore	7.0	4.4
Lower-48 offshore	0.7	0.4
Alaska	0.4	0.2
Total	8.1	5.0

SOURCE: Kuuskraa et al. (1989).

Crude oils with API gravity between 10° and 20° (or with viscosity between 100 and 10,000 cp) are generally considered to be heavy. U.S. and world resources of heavy oil are shown in Tables C-4 and C-5, respectively. They constitute an important but only partly developed resource. Venezuela, the Middle East, the United States, and the USSR have 50, 30, 10, and 5 percent of the world's heavy oil, respectively. The majority of the U.S. reservoirs and fields (with over half of the original oil in place) are in California, and there are important deposits in Alaska. The world's production of heavy oil is estimated at around 1.5 billion bbl/year, or around 7 percent of the world's total crude oil output. In the United States, heavy oil constitutes about 4 percent of total crude production (Riva, 1987).

U.S. and world resources of tar sands are shown in Table C-6. A tar sand (or oil sand) deposit (also known as a bitumen deposit) is defined as a hydrocarbon deposit having an in situ viscosity greater than 10,000 cp at reservoir conditions or an API gravity less than 10°. In cases where a deposit contains bitumen, heavy oil, and lighter oil, the deposit is considered a tar sand if the majority of the hydrocarbon is not mobile at reservoir conditions (Kuuskraa, 1985). In the United States, tar sand deposits are found in a number of states but the main deposit of about 12 billion bbl is found in Utah with speculative resources in that state amounting to about another 20 billion bbl (IOCC, 1984). The resource in Alaska of 10 billion bbl is judged to be speculative. Other bitumen resources exist in Alabama,

TABLE C-4 U.S. Resources of Heavy Oil[a] (billion bbl)

	Kuuskraa et al. (1989)		Riva (1987)		
	Original Oil in Place[b]	Total Recoverable	Proved Reserves	Undiscovered Resources	Total Recoverable
Large Reservoirs[c]	80				
Small Reservoirs[c]	20				
Total	100	34[d]	18	24	52[d]

[a]Heavy oil is defined as crude oil with an API gravity of between 10° and 20° or a viscosity between 100 and 10,000 cp. By the same standards, light oil is defined as having an API gravity between 20° and 45°+ or a viscosity between 1- and 100 centipoises (Kuuskraa et al., 1989).

[b]Original oil in place is the estimated gross quantity of oil in a known reservoir prior to any production. It is independent of economic and operating considerations governing the recovery of the oil from the reservoir.

[c]Twenty million barrels of original oil in place is used to distinguish large from small reservoirs (Kuuskraa et al., 1989).

[d]Includes cumulative production to date of about 10 billion bbl.

TABLE C-5 World Resources of Heavy Oil (billion bbl)

	Proved Reserves	Undiscovered Resources	Total Recoverable
North America	23.0	30.1	64.8
Central and South America	279.5	16.2	308.5
Western Europe	8.0	0.2	9.0
USSR and Eastern Europe	7.2	20.6	33.1
Africa	3.6	0.6	4.6
Middle East	115.4	22.1	169.0
Far East and Oceania	12.9	3.6	19.1
World total	449.6	93.4	608.1[a]

[a]Includes cumulative production to date of over 65 billion bbl.

SOURCE: Riva (1987).

TABLE C-6 U.S. and World Resources of Tar Sands[a] (billion bbl)

	Measured Resources[b]	Speculative Resources[c]	In-Place Resources
United States	21.6[d]	41.1[d]	62.7[d]
Canada			1700[e]
Venezuela			692[f]
World Total			4000[e]

[a]A tar sand (or bitumen) deposit is defined as a hydrocarbon deposit having an in situ viscosity greater than 10,000 cp at reservoir conditions or an API gravity less than 10°. In cases where a deposit contains tar (bitumen), heavy oil, and lighter oil, the deposit is considered a tar sand if the majority of the hydrocarbon is not mobile at reservoir conditions (Kuuskraa et al., 1989).

[b]Measured resources are defined as that part of the resource that can be *deduced* to exist from well control, using core analysis and geology.

[c]Speculative resources are defined as that part of the resource that is *presumed* to exist from reported tar shows on drillers' logs and other geological interpretation.

[d]Kuuskraa et al. (1985).

[e]Information supplied by the Alberta Oil Sands and Tar Research Authority of Canada.

[f]Considine (1977).

Texas, California, Kentucky, and several other states. Although the U.S. resource base is not as extensive as in other areas in the world, it is nevertheless significant and could support a 20-year oil production of several million barrels per day.

Globally, tar sand deposits are distributed throughout the world, but the largest are in the Athabasca region of Western Canada (over 40 percent of an estimated 4 trillion bbl of oil in place) and in the Orinocco Belt of Venezuela. Significant exploitation of this resource currently occurs only in Canada, where it accounts for about 20 percent of total crude oil production.

U.S. and world resources of oil shale are shown in Table C-7. Oil shale is sedimentary rock containing solid organic matter that when heated to around 500°C yields hydrocarbons and other solid products. Although oil shales are found in many places throughout the world, nearly 60 percent of the world's potentially recoverable resources are concentrated in the United States. The second largest resource (about 30 percent of the worldwide recoverable resources) is in Brazil. In the United States the western oil shale deposits account for about 90 percent of the resource base. They are also richer with thicker deposits than the eastern shales, yielding between 20 and 40 U.S. gallons of liquid feedstocks per short ton of raw shale processed. Commercial exploitation of oil shale resources is very limited at the present time (Riva, 1987).

TABLE C-7 U.S. and World Resources of Oil Shale

	Recoverable Resources (billion barrels of oil[a])
United States	627
Western shales	
Piceance basin (Colorado)	380
Uinta basin (Utah)	51
Other basins	131
Eastern Shales	
Kentucky, Indiana, Ohio	65
South America (Brazil)	300
USSR	42
Africa (Zaire)	38
World total	1007

[a]Recovery factor = 37.5 percent of estimated in-place resource.

SOURCE: Riva (1987).

TABLE C-8 U.S. Natural Gas Resources at Year End 1986 (trillion cubic feet)[a]

	Existing Technology and Efficiency[b]		Advanced Technology and Efficiency[c]	
	<$3.00/Mcf	<$5.00/Mcf	<$3.00/Mcf	<$5.00/Mcf
Proved reserves	166 (162)	166 (187)	166 (163)	179 (187)
Reserve growth[d]	197 (197)	226 (226)	313 (N/A)	483 (N/A)
Undiscovered	144 (144)	233 (233)	202 (N/A)	338 (N/A)
Low permeability	70 (79)	119 (86)	130 (245)	300 (275)
Coalbed methane	8 (13)	12 (26)	40 (51)	90 (65)
Shale	10 (17)	15 (20)	30 (37)	40 (46)
Total	595 (612)	771 (778)	881	1430

[a] Numbers in parentheses are from ICF Resources (Kuuskraa et al., 1989). Note: Mcf = thousand cubic feet.
[b] DOE (1986 dollars).
[c] AAPG (1986 dollars).
[d] Includes inferred reserves, new pool, and reserve growth from gas and gas-associated oil reservoirs.

SOURCE: Kuuskraa et al. (1989); AAPG (1989b).

TABLE C-9 World Resources of Natural Gas (trillion cubic feet)

	Proved Reserves	
	Oil & Gas Journal	*World Oil*
North America	361.7	357.6
Central and South America	150.1	156.6
Western Europe	218.1	230.5
USSR and Eastern Europe	1479.3	1468.5
Africa	248.7	199.2
Middle East	1084.1	1166.5
Far East and Oceania	255.5	275.3
World total	3797.5	3854.2

SOURCE: EIA (1988).

U.S. and world resources of natural gas are shown in Tables C-8 and C-9, respectively. U.S. proved reserves are currently around 200 Tcf. This accounts for about 5 percent of the world's total, compared to nearly 40 percent for the USSR and 30 percent for the Middle East. U.S. inferred reserves and undiscovered resources are each estimated at over two times proved reserves, and expectations are high that a significant portion of these reserves will in fact be discovered and economically developed (BEG, 1988). The U.S. endowment of natural gas has been estimated to be in the range of 1500 Tcf (placing it third after the USSR and the Middle East), of which about 45 percent is thought to have been already produced (Riva, 1987). As noted earlier, recent reviews of the resource base suggest that this endowment could in fact be substantially larger. Current annual U.S. consumption of natural gas is around 18 Tcf, of which around 5 percent is imported. World annual consumption is around 65 Tcf (EIA, 1987b, 1989a).

U.S. and world resources of coal are shown in Table C-10. The United States has the world's most extensive minable deposits of anthracite and bituminous coal (around 35 percent of the total), followed by the USSR and China. These three countries account for about 60 percent of the recoverable resources in the world (EIA, 1987b).

Biomass resources have not been estimated with any certainty, in part because there are as yet no established criteria for defining the economic recoverability of fuels from such resources.

TABLE C-10 U.S. and World Resources of Coal (billion short tons)

	Anthracite and Bituminous	Lignite	Total Recoverable
United States	254.8	36.0	290.8
Other North America	7.0	2.7	9.7
Central and South America	5.7		5.7
Western Europe	35.5	64.2	99.7
USSR and Eastern Europe	201.5	153.4	354.9
Africa	71.3		71.3
Middle East	0.2		0.2
Far East and Oceania	143.7	42.6	186.3
World total	719.7	298.9	1018.6

SOURCE: EIA (1987b).

D

Cost Analysis Methods

STRUCTURE OF ANALYSIS

Most of the structure of the cost analysis is explained in Chapter 3, but a few details are further elaborated here. Cost estimates are based on the best available data in the literature. Obviously, there could be plant-specific variations in costs for different fuels and changed future circumstances. However, this analysis gives a reasonable estimate of relative costs for the purposes of ranking the various processes as to economic attractiveness.

With regard to the economic assumptions delineated in Chapter 3, prices for the various energy and nonenergy feedstocks are expressed, where appropriate, as a function of the prevailing crude oil price. For each technology, the cost per equivalent barrel of oil, measured in 1988 dollars, is calculated as the summary cost measure based on the assumed oil price environment. This calculation begins with the crude oil price, here defined as the average price of U.S. imported crude oil. The crude oil price implies prices of energy and nonenergy inputs to the various production processes. In particular, prices of natural gas, electricity, and corn (as a feedstock) are assumed to increase with crude oil prices (see Table D-1).

In calculating cost per equivalent barrel of oil, per-gallon product costs are first calculated by adding together feedstock costs, operation and maintenance (O&M) (energy plus nonenergy) costs, and annual capital charge and subtracting by-product credits, all on a per-gallon basis. Per-gallon feedstock costs are calculated by multiplying feedstock quantity per gallon times the price per unit of the feedstock. Energy O&M costs are calculated in the same manner. Nonenergy O&M costs are directly added. The per-gallon capital charge is calculated by dividing the annual capital cost by the annual production of the product. The annual capital cost is simply the

APPENDIX D

TABLE D-1 Economic Assumptions Used in the Cost Estimates

Natural Gas Price[a] ($/Mcf):[b]	$3.91 + 0.05857 \times (P_{oil} - 28)$
Corn price ($/bu):	$2.500 + 0.01786 \times (P_{oil} - 28)$
Corn by-product price ($/bu):	$1.200 + 0.01136 \times (P_{oil} - 28)$
Electricity price (bought) ($/kWh):[c]	$0.049 + 0.00020 \times (P_{oil} - 28)$
Gasoline refining (credit) ($/bbl):	$7.000 + 0.18182 \times (P_{oil} - 28)$
Gasoline distribution/marketing (credit) ($/bbl):	$4.000 + 0.02 \times (P_{oil} - 28)$
Distribution/marketing for product ($/bbl):	$4 \times EQ + 0.02 (P_{eq} - 28)$
Coal price ($/ton):	$38.00
Oil shale feedstock ($/ton):	$6.02
Tar sands feedstock ($/ton):	$8.00
Wood price ($/dry ton):	$32.40
Oxygen ($/ton):	$57.05
Form coke ($/ton):	$100.00
Capital charge rate (%/year):	16.0% and 24%
Consumer discount rate:	10% and 15%
Automobile efficiency:	27 mpg
Natural gas delivery cost:	$0.94/Tcf

NOTE: P_{eq} = cost per equivalent barrel of oil, net of the additional vehicle cost; P_{oil} = price of petroleum in 1988 dollars; EQ = 1.8 for methanol; 1.5 for ethanol.

[a]In the present analysis, natural gas prices are not allowed to exceed $5.00/Mcf (1988 dollars) because coal gasification becomes competitive above this price.

[b]EIA Base Case forecast for the year 2000 has $28/barrel petroleum and $3.91/Mcf for natural gas; High World Oil Price Case has $35/barrel petroleum and $4.32/Mcf natural gas. Hence, ($4.32 - $3.91)/7 = 0.05857 is the slope in the equation for projecting gas price as a function of petroleum price.

[c]EIA Base Case forecast of $0.04896/kWh for the year 2000 and High World Oil Price Case of $0.05038/kWh for electricity. Hence, the slope in the equation is $(0.05038 - 0.04896)/7 = 0.0002$.

investment cost multiplied by the annual capital charge factor. By-product credits are calculated as a price per unit multiplied by the by-product quantity produced (per gallon of primary product) or are directly estimated.

SPECIFIC FACTORS

Table D-1 summarizes the specific input factor assumptions for the economic analysis of the various fuel production technologies. The relationships between crude oil price and other energy prices were calibrated from the Base Case Forecasts and the High World Oil Price Forecasts published by the Energy Information Administration (EIA) in its 1989 Annual Energy Outlook (EIA, 1989a; see Table D-2). The EIA Low World Oil Price Scenarios were not used for the calibrations. The Base Case Forecasts and High World Oil Price Forecasts assumed for the year 2000 are $28/barrel and $35/barrel (1988 dollars), respectively. The EIA estimated prices of coal delivered to industrial users at $38/short ton under either of these scenarios. The committee's analysis used this price for all world oil prices.

The EIA estimated wellhead natural gas prices as $3.91/Mcf in the Base Case and $4.32/Mcf in the High World Oil Price Case. Costs of gas delivered to industrial users were assumed to be $0.94/Mcf higher in both cases. In the present analysis this relationship between oil prices and wellhead natural gas prices is linearly extrapolated as oil prices increase, but natural

TABLE D-2 Prices of Energy in the Year 2000 from the Energy Information Administration's Forecasts in 1988 Dollars

	Year		
	1990	1995	2000
Low World Oil Price Case			
World oil price ($/bbl)[a]	12.89	16.70	1.70
Natural gas ($/Mcf)[b]	1.69	2.49	3.52
Coal ($/ton)[c]	23.76	24.61	25.42
Electricity ($/kWh)[d]	0.0487	0.0461	0.0472
Base Case Forecasts			
World oil price ($/bbl)	15.00	20.60	28.00
Natural gas ($/Mcf)	1.75	2.80	3.91
Coal ($/ton)	24.00	24.95	25.87
Electricity ($/kWh)	0.0489	0.04704	0.04896
High World Oil Price Case			
World oil price ($/bbl)	18.00	24.40	35.00
Natural gas ($/Mcf)	1.85	3.90	4.32
Coal ($/ton)	24.34	25.41	26.14
Electricity ($/kWh)	0.049	0.04798	0.05038

[a] Cost of imported crude oil to U.S. refiners.
[b] Average wellhead price.
[c] Mine-mouth price.
[d] Price for industrial users.

SOURCE: EIA (1989a).

gas prices are not allowed to exceed $5.00/Mcf (1988 dollars) because coal gasification becomes competitive above this price. Natural gas used as a feedstock is assumed to reflect the wellhead price, reflecting a conversion plant operating on the Gulf Coast near to both natural-gas-producing wells and the distribution system for final products. Natural gas used as an operating cost, but not as a feedstock, is assumed to reflect a cost equivalent to that facing average industrial users.

The EIA estimated average electricity price delivered to industrial users in the year 2000 as $0.04896/kWh and $0.05038/kWh in the Base and High World Oil Price cases, respectively. A linear relationship between oil and electricity prices was derived for the present analysis from these points. Corn prices and corn by-product prices were also assumed to be linearly related to oil prices, and prices for oil shale feedstock, wood, and oxygen were assumed. The relationship between corn price and crude oil price was based on the assumption that every bushel of corn requires three-quarters of a gallon of oil in farming.

Refining costs of gasoline were assumed to increase with the price of crude oil. The relationship was calibrated so that the spread between gasoline and crude oil price would be $7/barrel when the crude oil price was $28/barrel and would increase by $2/barrel for every $11/barrel increase in the crude oil price. This gave a refining credit for methanol, ethanol, and compressed natural gas (CNG).

Distribution and marketing cost (per actual barrel) was assumed to be $4/barrel of product when crude oil was $28/barrel. The same cost per actual barrel (or per gallon) was applied to gasoline, methanol, and ethanol. Distribution costs were also related to the value of the product being distributed. Distribution costs were assumed to increase by $0.02/barrel for every $1/barrel increase in product costs. This increase was meant to reflect additional inventory costs of the higher-valued product. The same relationship was applied to all methanol, gasoline, and ethanol.

An equivalency factor of 1.80 was used to convert a given volume of methanol to the volume of gasoline that would give the same work output as a methanol-fueled automobile. This factor arose from the assumption that for a specially designed methanol vehicle there would be a 10 to 18 percent energy efficiency advantage over today's gasoline-fueled automobiles. This energy advantage could vary from approximately zero, for first-generation dual-fueled automobiles to 30 percent or more for future, highly optimized, methanol-fueled systems. For zero efficiency difference the equivalency factor would be from 2.02 to 2.06 (2.06 is used in the present study) (based on the heat of combustion in American Petroleum Institute Report 4621 [1976]). For 30 percent efficiency advantage the equivalency factor would be from 1.55 to 1.57 (1.57 is used in the present study).

Distribution and marketing cost for CNG was separately analyzed, since the compressor station and the distribution station would typically be one unit. Thus, capital cost for this technology was the cost of a single filling station that would deliver a gasoline equivalent of 1200 gal/day and would use 0.12 MMBtu of natural gas per equivalent gallon of gasoline. Annual operating and maintenance costs (nonenergy) were estimated as 7 percent of the original investment cost. No separate additional distribution and marketing costs were assessed.

For two fuels, methanol and CNG, the capital cost for new cars was anticipated to be greater than for the other fuels. For these fuels this additional capital cost was translated to an equivalent per-mile additional operating cost, where the per-mile equivalent cost was derived so as to give the same discounted present value of costs to the consumer as would the one-time additional capital cost. For this calculation it was assumed that a typical car was driven 15,000 miles per year over the course of 7 years. The per-mile equivalent cost was then translated to a per-gallon cost by multiplying by the average fuel efficiency of new cars, assumed to be 27 mpg. The specific equation to calculate the per-mile equivalent cost is as follows:

$$K = \frac{15,000 \times C_{eq} \, [1 - 1/(1 + r)^7] \, (1 + r)}{r},$$

where K is the additional capital cost of a new automobile, C_{eq} is the per-mile equivalent cost, and r is the consumer discount rate, assumed to be 10 percent annually (real). The right-hand side of the equation discounts the 7 years of annual costs (equal to $15,000 \times C_{eq}$ each year) back to the time that the car is purchased. The left-hand side is simply the additional purchase price. The equation can be solved for C_{eq}. The per-gallon equivalent cost is thus the product of C_{eq} and the fuel efficiency (in mpg) of future automobiles.

Discount Rate

A point of some uncertainty is the appropriate discount rate assumptions to use in analyzing the cost of capital-intensive projects, such as those considered in the study. One perspective is that the appropriate discount rate for a policy planning study should be equal to the estimated cost of capital facing corporations of the types that might consider investing in such projects. A second perspective is that the discount rate should be equal to the estimated typical hurdle rates for investments made by such corporations.

The first perspective has led to the 10 percent real discount rate used in the base case. An analysis of the average historical returns to equity capital

APPENDIX D

in financial markets and of the historical returns on physical capital invested in United States industry suggests a real cost of capital ranging between 8 and 12 percent. Variations in the estimate depend on the riskiness of an industry and the degree to which the risks are correlated with returns to the overall economy. The 10 percent figure has been selected as typifying that range.

The second perspective, that discount rates should be based on "typical" corporate hurdle rates, has led to the 15 percent real discount rate also used. The hurdle rate is the minimum estimated rate of return required by a corporation in order to approve a prospective investment. Hurdle rates are chosen within corporations so as to guide investment decision making. Since they are corporate policy instruments, hurdle rates can vary widely among corporations and can vary within a corporation based on the nature of the prospective investment. Thus, hurdle rates might well be higher than 15 percent, particularly for projects seen to be particularly risky. Hurdle rates can be lower than 15 percent, particularly if a project is not seen as being more risky than typical other investments.

The perspective based on typical hurdle rates leads to a higher discount rate because a firm's hurdle rates typically exceed its costs of capital. Reasons for this difference between discount rate and cost of capital vary among firms. Hurdle rates are often applied not to the expected value of revenues and costs but to estimated values of revenues and costs based on some risks but ignoring others that cannot be reasonably quantified. Hurdle rates exceeding the cost of capital can compensate for the inability to include some of the downside risks. However, evaluations based on estimations of expected value already account for the important quantifiable risks. In addition, some firms increase the hurdle rate to compensate for the natural tendency for project advocates within a corporation to present the economics of their projects using rather favorable or optimistic assumptions. Finally, increased hurdle rates can compensate for a "winner's curse" phenomenon in which random variations in cost estimates lead the given project to be pursued dominantly by those groups that happen to most overestimate its profitability and to be rejected by those that happen to most underestimate its profitability.

The study used 10 and 15 percent discount rates. Summary discussions make primary use of the 10 percent cases, which past studies indicate are typical of industrial returns under relatively stable conditions. Also, for the purpose of government R&D planning, the Office of Management and Budget specifies a 10 percent cost of capital based on considerations detailed above. Early application of any of these technologies will entail risks because of uncertainties in technology and oil price fluctuations and, in the case of alternative fuels, consumer acceptance. These pioneer plants would require either a high hurdle rate or risk reduction mechanisms by the government.

Without such mechanisms the construction of such pioneer plants would wait for higher calculated returns on investment or reduction of perceived risks. For a mature and developed industry the costs of capital should approximate those typical for refinery resid conversion processes plus a premium to reflect the risk of lower than anticipated crude oil prices. This would suggest a 15 percent cost of capital.

Annual Capital Charge Factors

Annual capital charge factors are calculated so that the net present value of the stream of capital charges, after taxes, is just equal to the initial investment cost, using the cost of capital as the discount rate. A project life is chosen as 20 years, plus the construction time. It is assumed that the median of investment costs is incurred 2 years before the middle of the first year of plant operation or, equivalently, 1.5 years before the project begins operating. (More precisely, it is assumed that investment costs are spread over time so that the present value of these expenditures, discounted to a point 1.5 years before the project begins operating, is equal to the total investment cost.)

In developing these factors it is assumed that the various processes face tax rules consistent with the current tax laws. In particular, it is assumed that the corporation pays a 34 percent tax rate on profits and that the investment is depreciated over time using a 10-year, double-declining balance depreciation schedule. Depreciation allowances depend on the nominal value of historical investment costs and are not adjusted upward for inflation. The tax deductibility of interest payments is incorporated implicitly in the analysis through the use of the after-tax cost of capital. (When present-value calculations are conducted using the after-tax cost of capital, no additional deductions for interest payments should be included explicitly.)

Plant Investment Costs

Plant investment costs are based on the committee's estimates of costs that might characterize a developed industry. Even though these technologies in many cases have not been fully demonstrated and have not been commercialized, the committee has not based its estimates on costs of the first demonstration or pioneer plant. In these technology assessments the committee has tried to assess the median value of the capital costs for a project. No explicit probabilistic analysis was conducted by the committee.

Capital investment cost estimates have been adjusted to 1988 constant dollars to be consistent with the other economic factors. Table D-3 displays estimates of these investment costs (denoted as capital or capital/capacity) for each technology. In calculating investment costs an attempt has been made to include all of the relevant costs of the investment, including those for process (onsite) and offsite construction, infrastructure development (where needed), planning, environmental compliance, site preparation, and contingencies.

Plant capacities have been chosen based on the economically efficient sizes for the various technologies. The committee has attempted to standardize the capacities across various technologies when economies of scale appeared to be important in influencing overall costs; however, all capacities are not standardized to one common level. Table D-3 displays the capacities assumed for the various technologies. Capacities are expressed both in terms of actual barrels per day and in terms of oil equivalent barrels per day.

To provide some comparability among the technologies, it is useful to calculate the ratio of investment costs to capacities to obtain an investment cost per daily barrel of capacity. The ratio can be based on either actual barrels or capacity of oil equivalent barrels. Both of these ratios are provided in the output tables.

Figure D-1 shows the investment cost per oil equivalent daily barrel of capacity for each technology. In what follows, this ratio will be referred to as the "per-barrel investment cost." These figures can be interpreted as the

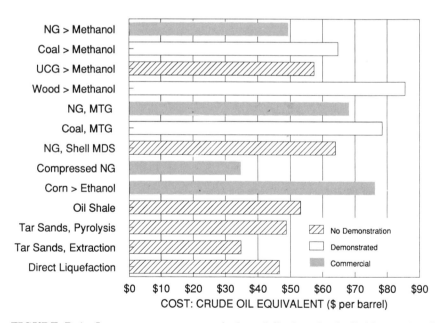

FIGURE D-1 Investment cost per equivalent daily barrel of oil (thousands of dollars at 1988 prices). New estimated capital cost for coal-to-methanol could be 40 percent lower.

initial investment cost for each 1 barrel per day of capacity to produce a fuel that would substitute for 1 barrel per day of oil. As can be seen, there is wide variability in the estimates of the per-barrel investment cost. In general, these per-barrel investment costs exceed the $10,000 to $20,000 range typical of investments for crude oil extraction in the United States, ranging from a low of $20,000 for production facilities for methanol from natural gas to a high of $102,000 for gasoline produced from coal using methanol-to-gasoline conversion processes. While the CNG investment cost is shown as lower than the range cited, these figures include the investment cost only for a CNG delivery station and do not include the additional costs of the vehicles themselves or the investment cost of natural gas wells.

COST ESTIMATES FOR THE VARIOUS TECHNOLOGIES

Tables D-3 to D-8 show cost estimates for the various technologies. Table D-1 provides a detailed statement of the various assumptions for each

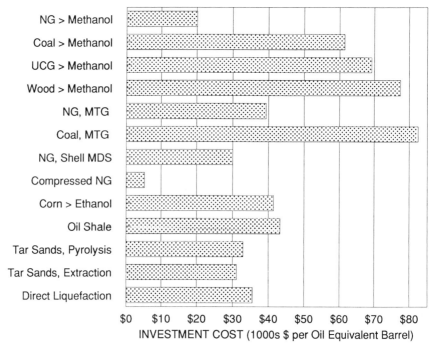

FIGURE D-2 Estimated costs of alternative fuels (15 percent discounted cash flow, endogenous price calculation).

technology and Table D-3 shows the resulting costs on a disaggregated basis. Table D-4 shows results based on 10 and 15 percent real discount rates and the endogenous determination of energy prices.

Tables D-4 to D-8 show costs under a number of combinations of the various inputs. These sensitivity studies show variations of costs with respect to the discount rate, the crude oil price, and the natural gas price function. Additional details on source data used in the economic analysis are provided in Table D-9.

Graphs showing cost estimates and their components are presented in Chapter 3. Figure D-2 shows cost data similar to the data in the body of this report for the 15 percent discount rate.

Plant capacities for the coal to methanol, coal to methanol and gasoline, and the natural gas to methanol to gasoline technologies were adjusted by the committee to the larger plant scale of 50,000 bbl/day oil equivalent. This scale is consistent with the scale used for the natural gas to methanol technology. A conservative scaling factor of 0.85 was used.

However, the scaling exponent could be as low as 0.7 if the same economies of scale apply based on Bechtel's recent analyses of natural gas to methanol plant investments. In addition to estimating the cost of an 80,000-bbl/day methanol plant (in the California Fuel Methanol Study [1989]), the cost for a 20,000-bbl/day plant was estimated.

If the investments for the 50,000-bbl/day plants using the above three technologies were estimated using a 0.7 rather than 0.85 scaling exponent, the capital investment would be 15 to 20 percent lower. The cost estimates are summarized as follows:

Process:	Coal to Methanol		Coal, MTG[a]		Natural Gas, MTG[b]	
Scaling exponent	0.85	0.7	0.85	0.7	0.85	0.7
Capital investment[c]	34,122	26,800	82,460	69,900	39,320	33,320
Cost ($/barrel)[d]	53	45	62	55	60	57

[a] Coal as feedstock, with Mobile methanol-to-gasoline (MTG) process.
[b] Natural gas as feedstock, with the MTG process.
[c] Capital investment is dollars per actual barrel/day.
[d] Cost is per crude oil equivalent barrel, 10 percent discount rate, and endogenous price calculation.

TABLE D-3 Base Case Analysis (10 percent discount rate; energy prices endogenous)

Cost Factors	NG/ Methanol	Coal/ Methanol	Underground Coal Gasification/ Methanol	Wood/ Methanol	NG/ MTG	Coal/ MTG	NG/ Shell MDS	NG/ CNG	Corn/ Ethanol	Oil Shale/ Fluid Bed	Tar Sands/ Pyrolysis	Tar Sands/ Extraction	Coal/ Liquefaction
Gasoline Equivalency	1.80	1.80	1.80	1.80	1.00	1.00	1.00	1.00	1.50	1.00	1.00	1.00	1.00
Capacity (bbl/day, oil equivalent)	43,907	50,000	9,970	308	50,000	50,000	10,000	29	9,660	54,560	22,500	25,389	80,944
Capacity (bbl/day, actual)	79,033	90,000	17,946	554	50,000	50,000	10,000	29	14,490	54,560	22,500	25,389	80,944
Capacity Factor	95.0%	90.4%	90.4%	90.4%	90.4%	90.4%	90.4%	100.0%	90.4%	90.4%	90.4%	90.0%	90.0%
Average Production (bbl/day, actual)	75,081	81,370	16,225	501	45,205	45,205	9041	29	13,101	49,328	20,342	2,2850	72,850
Capital ($MM)	$883	$3,071	$690	$24	$1,966	$4,123	$300	$0.15	$400	$2,360	$740	$790	$2,871
Capital/Capacity	$11,173	$34,122	$38,449	$43,056	$39,320	$82,460	$30,000	$5,250	$27,605	$43,255	$32,889	$31,116	$35,469
Capital/Capacity (oil equivalent)	$20,111	$61,420	$69,208	$77,500	$39,320	$82,460	$30,000	$5,250	$41,408	$43,255	$32,889	$31,116	$35,469
Vehicle Capital Cost	$200.00	3.11 $200.00	$200.00	$200.00				$1,000.00					
Operating: Feedstocks													
Coal (lb/gal)		11.30				24.00							13.76
(lb/bbl)	0.00	474.60	0.00	0.00	0.00	1,008.00	0.00	0.00	0.00	0.00	0.00	0.00	578.00
Natural gas (Mcf/gal)	0.095				0.22		0.24	0.12					
Wellhead (Mcf/bbl)	4.00	0.00	0.00	0.00	9.24	0.00	10.00	5.04	0.00	0.00	0.00	0.00	0.20
Delivered (Mcf/bbl)													
Vacuum residue													
Wood (lb/gal)				13.40									
Corn (bu/gal)									0.40				
Oil shale (ton/bbl)	0.00	0.00	0.00	0.00	0.00	0.00	0.00	0.00	0.00	0.03 1.33	0.00	0.00	0.00
Tar sands (ton/gal)											0.00 0.04	0.00	
(ton/bbl)	0.00	0.00	0.00	0.00	0.00	0.00	0.00	0.00	0.00	0.00	1.50	0.00	0.00
Oxygen (lb/gal)				3.35									

Nonenergy	$0.06	$0.16	$0.18	$0.20	$0.18	$0.38	$0.14	$0.15	$0.16	$0.21	$0.17	$0.33	$0.25
Percentage of investment		7.0%		7.0%	7.0%	7.0%	7.0%		7.0%				
Cost per barrel	$2.34		$7.56					$6.30	$1.26	$9.00	$7.00	$14.05	$10.36
Power	$0.00	$0.00	$0.00	$0.03	$0.00	$0.00	$0.00	$0.03	$0.06	$0.01	$0.00	$0.00	$0.12
kWh/gal								0.67	1.00				
MW				0.50						9.80	0.02		339.72
Natural gas (Mcf/gal)										0.02	1.00		
(Mcf/bbl)	0.00	0.00	0.00	0.00	0.00	0.00	0.00	0.00	0.00	0.89		0.00	0.00
Coal (lb/gal)									3.30				
(lb/bbl)									138.60				
By-products													
Ammonia, sulfur, etc.										($0.02)		($0.01)	($0.03)
Fuel gas, form coke									($0.63)				
Animal feed													
Product Cost ($/gal)													
Feed	$0.465	$0.215	$0.000	$0.313	$1.100	$0.456	$1.190	$0.507	$1.270	$0.190	$0.286	$0.000	$0.287
O&M, nonenergy	$0.056	$0.156	$0.180	$0.197	$0.180	$0.377	$0.137	$0.150	$0.156	$0.214	$0.167	$0.335	$0.247
O&M, energy	$0.000	$0.000	$0.000	$0.029	$0.000	$0.000	$0.000	$0.034	$0.119	$0.126	$0.133	$0.000	$0.122
Capital charge	$0.124	$0.398	$0.449	$0.502	$0.459	$0.962	$0.350	$0.055	$0.322	$0.505	$0.384	$0.365	$0.416
By-products	$0.000	$0.000	$0.000	$0.000	$0.000	$0.000	$0.000	$0.000	($0.635)	($0.017)	$0.000	($0.010)	($0.028)
TOTAL	$0.65	$0.77	$0.63	$1.04	$1.74	$1.79	$1.68	$0.75	$1.23	$1.02	$0.97	$0.69	$1.04
Product Cost ($/bbl)													
Feed	$19.55	$9.02	$0.00	$13.13	$46.20	$19.15	$50.00	$21.29	$53.33	$8.00	$12.00	$0.00	$12.07
O&M, nonenergy	$2.34	$6.54	$7.56	$8.26	$7.54	$15.81	$5.75	$6.30	$6.55	$9.00	$7.00	$14.05	$10.36
O&M, energy	$0.00	$0.00	$0.00	$1.23	$0.00	$0.00	$0.00	$1.41	$5.01	$5.31	$5.59	$0.00	$5.14
Capital charge	$5.21	$16.72	$18.84	$21.10	$19.27	$40.41	$14.70	$2.33	$13.53	$21.20	$16.12	$15.32	$17.46
By-products	$0.00	$0.00	$0.00	$0.00	$0.00	$0.00	$0.00	$0.00	($26.67)	($0.70)	$0.00	($0.41)	($1.16)
TOTAL	$27.10	$32.28	$26.40	$43.72	$73.01	$75.37	$70.45	$31.32	$51.76	$42.81	$40.71	$28.96	$43.86

TABLE D-3 Continues

TABLE D-3 Continued

			Feed/Fuel or Process									
Cost Factors	NG/ Methanol	Coal/ Methanol	Underground Coal Gasification/ Methanol	Wood/ Methanol	NG/ MTG	Coal/ MTG	NG/ Shell MDS	NG/ CNG	Corn/ Ethanol	Oil Shale/ Fluid Bed	Tar Sands/ Pyrolysis	Tar Sands/ Extraction	Coal/ Liquefaction
Product Cost ($/equivalent bbl)													
Feed	$35.19	$16.23	$0.00	$23.64	$46.20	$19.15	$50.00	$21.29	$80.00	$8.00	$12.00	$0.00	$12.07
O&M total	$4.20	$11.78	$13.61	$17.08	$7.54	$15.81	$5.75	$7.71	$17.35	$14.31	$12.59	$14.05	$15.50
Capital charge	$9.38	$30.10	$33.91	$37.98	$19.27	$40.41	$14.70	$2.33	$20.29	$21.20	$16.12	$15.32	$17.46
By-products	$0.00	$0.00	$0.00	$0.00	$0.00	$0.00	$0.00	$0.00	($40.00)	($0.70)	$0.00	($0.41)	($1.16)
TOTAL	$48.77	$58.11	$47.52	$78.70	$73.01	$75.37	$70.45	$31.32	$77.64	$42.81	$40.71	$28.96	$43.86
Gasoline Equivalent Costs													
$/gal	$1.161	$1.383	$1.131	$1.874	$1.738	$1.795	$1.677	$0.746	$1.848	$1.019	$0.969	$0.689	$1.044
$/bbl	$48.771	$58.107	$47.520	$78.696	$73.008	$75.372	$70.454	$31.322	$77.637	$42.808	$40.710	$28.957	$43.863
Distribution/Marketing ($/bbl)	$7.478	$7.636	$7.456	$7.984	$4.643	$4.683	$4.600	$0.746	$6.755				
Refining Credit	($10.037)	($11.473)	($9.844)	($14.640)	($12.847)	($13.211)	($12.454)	($7.974)	($13.867)			($1.400)	($5.900)
Gasoline Distribution/Marketing Total	($4.334)	($4.492)	($4.313)	($4.840)	($4.643)	($4.683)	($4.600)	($4.107)	($4.755)				
	($6.89)	($8.33)	($6.70)	($11.50)	($12.85)	($13.21)	($12.45)	($12.08)	($11.87)	$0.00	$0.00	($1.40)	($5.90)
TOTAL COST ($/bbl)	$41.88	$49.78	$40.82	$67.20	$60.16	$62.16	$58.00	$19.24	$65.77	$42.81	$40.71	$27.56	$37.96
Vehicle Cost ($/gal additional)	$0.07	$0.07	$0.07	$0.07	$0.00	$0.00	$0.00	$0.34	$0.00	$0.00	$0.00	$0.00	$0.00
($/bbl additional)	$2.82	$2.82	$2.82	$2.82	$0.00	$0.00	$0.00	$14.12	$0.00	$0.00	$0.00	$0.00	$0.00
COST ($/oil bbl equivalent)	$44.70	$52.60	$43.64	$70.02	$60.16	$62.16	$58.00	$33.36	$65.77	$42.81	$40.71	$27.56	$37.96
Prices of Feedstocks													
Crude oil ($/bbl)	$44.70	$52.60	$43.64	$70.02	$60.16	$62.16	$58.00	$33.36	$65.77	$42.81	$40.71	$27.56	$37.96
Natural gas ($/Mcf)	$4.89	$5.00	$4.83	$5.00	$5.00	$5.00	$5.00	$4.22	$5.00	$4.78	$4.65	$3.88	$4.49
Corn ($/bu)	$2.80	$2.94	$2.78	$3.25	$3.07	$3.11	$3.04	$2.60	$3.17	$2.76	$2.73	$2.49	$2.68
Electricity ($/kWh)	$0.052	$0.054	$0.052	$0.057	$0.055	$0.056	$0.055	$0.050	$0.057	$0.052	$0.052	$0.049	$0.051
Electricity (sold)	$0.042	$0.044	$0.042	$0.047	$0.045	$0.046	$0.045	$0.040	$0.047	$0.042	$0.042	$0.039	$0.041

TABLE D-4 Costs of Energy Conversion Technologies for 10 and 15 Percent Discount Rates (endogenous price calculations)

Cost Factors	NG/ Methanol	Coal/ Methanol	Underground Coal Gasification/ Methanol	Wood/ Methanol	NG/ MTG	Coal/ MTG	NG/ Shell MDS	NG/ CNG	Corn/ Ethanol	Oil Shale/ Fluid Bed	Tar Sands/ Pyrolysis	Tar Sands/ Extraction	Coal/ Liquefaction
10 PERCENT DISCOUNT RATE													
Product Cost ($/gal)													
Feed	$0.465	$0.215	$0.000	$0.313	$1.100	$0.456	$1.190	$0.507	$1.270	$0.190	$0.286	$0.000	$0.287
O&M, nonenergy	$0.056	$0.156	$0.180	$0.197	$0.180	$0.377	$0.137	$0.150	$0.156	$0.214	$0.167	$0.335	$0.247
O&M, energy	$0.000	$0.000	$0.000	$0.029	$0.000	$0.000	$0.000	$0.034	$0.119	$0.126	$0.133	$0.000	$0.122
Capital charge	$0.124	$0.398	$0.449	$0.502	$0.459	$0.962	$0.350	$0.055	$0.322	$0.505	$0.384	$0.365	$0.416
By-products	$0.000	$0.000	$0.000	$0.000	$0.000	$0.000	$0.000	$0.000	($0.635)	($0.017)	$0.000	($0.010)	($0.028)
TOTAL	$0.65	$0.77	$0.63	$1.04	$1.74	$1.79	$1.68	$0.75	$1.23	$1.02	$0.97	$0.69	$1.04
Product Cost ($/bbl)													
Feed	$19.55	$9.02	$0.00	$13.13	$46.20	$19.15	$50.00	$21.29	$53.33	$8.00	$12.00	$0.00	$12.07
O&M, nonenergy	$2.34	$6.54	$7.56	$8.26	$7.54	$15.81	$5.75	$6.30	$6.55	$9.00	$7.00	$14.05	$10.36
O&M, energy	$0.00	$0.00	$0.00	$1.23	$0.00	$0.00	$0.00	$1.41	$5.01	$5.31	$5.59	$0.00	$5.14
Capital charge	$5.21	$16.72	$18.84	$21.10	$19.27	$40.41	$14.70	$2.33	$13.53	$21.20	$16.12	$15.32	$17.46
By-products	$0.00	$0.00	$0.00	$0.00	$0.00	$0.00	$0.00	$0.00	($26.67)	($0.70)	$0.00	($0.41)	($1.16)
TOTAL	$27.10	$32.28	$26.40	$43.72	$73.01	$75.37	$70.45	$31.32	$51.76	$42.81	$40.71	$28.96	$43.86
Product Cost ($/equivalent bbl)													
Feed	$35.19	$16.23	$0.00	$23.64	$46.20	$19.15	$50.00	$21.29	$80.00	$8.00	$12.00	$0.00	$12.07
O&M total	$4.20	$11.78	$13.61	$17.08	$7.54	$15.81	$5.75	$7.71	$17.35	$14.31	$12.59	$14.05	$15.50
Capital charge	$9.38	$30.10	$33.91	$37.98	$19.27	$40.41	$14.70	$2.33	$20.29	$21.20	$16.12	$15.32	$17.46
By-products	$0.00	$0.00	$0.00	$0.00	$0.00	$0.00	$0.00	$0.00	($40.00)	($0.70)	$0.00	($0.41)	($1.16)
TOTAL	$48.77	$58.11	$47.52	$78.70	$73.01	$75.37	$70.45	$31.32	$77.64	$42.81	$40.71	$28.96	$43.86

TABLE D-4 Continues

TABLE D-4 Continued

			Feed/Fuel or Process									
Cost Factors	NG/ Methanol	Coal/ Methanol	Underground Coal Gasification/ Methanol	Wood/ Methanol	NG/ MTG	Coal/ MTG	NG/ Shell MDS	NG/ CNG	Corn/ Ethanol	Oil Shale/ Fluid Bed	Tar Sands/ Pyrolysis	Tar Sands/ Extraction	Coal/ Liquefaction
Gasoline Equivalent Costs													
$/gal	$1.161	$1.383	$1.131	$1.874	$1.738	$1.795	$1.677	$0.746	$1.848	$1.019	$0.969	$0.689	$1.044
$/bbl	$48.771	$58.107	$47.520	$78.696	$73.008	$75.372	$70.454	$31.322	$77.637	$42.808	$40.710	$28.957	$43.863
Distribution/Marketing ($/bbl)	$7.478	$7.636	$7.456	$7.984	$4.643	$4.683	$4.600		$6.755				
Refining Credit	($10.037)	($11.473)	($9.844)	($14.640)	($12.847)	($13.211)	($12.454)	($7.974)	($13.867)		($1.400)		($5.900)
Gasoline Distribution/Marketing Total	($4.334)	($4.492)	($4.313)	($4.840)	($4.643)	($4.683)	($4.600)	($4.107)	($4.755)				
	($6.89)	($8.33)	($6.70)	($11.50)	($12.85)	($13.21)	($12.45)	($12.08)	($11.87)	$0.00	$0.00	($1.40)	($5.90)
TOTAL COST ($/bbl)	$41.88	$49.78	$40.82	$67.20	$60.16	$62.16	$58.00	$19.24	$65.77	$42.81	$40.71	$27.56	$37.96
Vehicle Cost ($/gal additional)	$0.07	$0.07	$0.07	$2.82	$0.00	$0.00	$0.00	$0.34	$0.00	$0.00	$0.00	$0.00	$0.00
($/bbl additional)	$2.82	$2.82	$2.82	$2.82	$0.00	$0.00	$0.00	$14.12	$0.00	$0.00	$0.00	$0.00	$0.00
COST ($/oil bbl equivalent)	$44.70	$52.60	$43.64	$70.02	$60.16	$62.16	$58.00	$33.36	$65.77	$42.81	$40.71	$27.56	$37.96
Prices of Feedstocks													
Crude oil ($/bbl)	$44.70	$52.60	$43.64	$70.02	$60.16	$52.16	$58.00	$33.36	$65.77	$42.81	$40.71	$27.56	$37.96
Natural gas ($/Mcf)	$4.89	$5.00	$4.83	$5.00	$5.00	$5.00	$5.00	$4.22	$5.00	$4.78	$4.65	$3.88	$4.49
Corn ($/bu)	$2.80	$2.94	$2.78	$3.25	$3.07	$3.11	$3.04	$2.60	$3.17	$2.76	$2.73	$2.49	$2.68
Electricity ($/kWh)	$0.052	$0.054	$0.052	$0.057	$0.055	$0.056	$0.055	$0.050	$0.057	$0.052	$0.052	$0.049	$0.051
15 PERCENT DISCOUNT RATE													
Product Cost ($/gal)													
Feed	$0.476	$0.215	$0.000	$0.313	$1.100	$0.456	$1.190	$0.516	$1.343	$0.190	$0.286	$0.000	$0.290
O&M, nonenergy	$0.056	$0.156	$0.180	$0.197	$0.180	$0.377	$0.137	$0.150	$0.156	$0.214	$0.167	$0.335	$0.247
O&M, energy	$0.000	$0.000	$0.000	$0.031	$0.000	$0.000	$0.000	$0.034	$0.121	$0.131	$0.141	$0.000	$0.126
Capital charge	$0.183	$0.588	$0.662	$0.742	$0.678	$1.421	$0.517	$0.082	$0.476	$0.745	$0.567	$0.539	$0.614
By-products	$0.000	$0.000	$0.000	$0.000	$0.000	$0.000	$0.000	$0.000	($0.671)	($0.017)	$0.000	($0.010)	($0.028)
TOTAL	$0.71	$0.96	$0.84	$1.28	$1.96	$2.25	$1.84	$0.78	$1.42	$1.26	$1.16	$0.86	$1.25

Product Cost ($/bbl)										
Feed	$20.00	$9.02	$0.00	$13.13	$46.20	$19.15	$50.00	$21.65	$56.41	$8.00	$12.00	$0.00	$12.17
O&M, nonenergy	$2.34	$6.54	$7.56	$8.26	$7.54	$15.81	$5.75	$6.30	$6.55	$9.00	$7.00	$14.05	$10.36
O&M, energy	$0.00	$0.00	$0.00	$1.30	$0.00	$0.00	$0.00	$1.42	$5.10	$5.52	$5.94	$0.00	$5.31
Capital charge	$7.69	$24.69	$27.82	$31.16	$28.46	$59.68	$21.71	$3.44	$19.98	$31.30	$23.80	$22.62	$25.79
By-products	$0.00	$0.00	$0.00	$0.00	$0.00	$0.00	$0.00	$0.00	($28.20)	($0.70)	$0.00	($0.41)	($1.16)
TOTAL	$30.03	$40.26	$35.38	$53.85	$82.20	$94.64	$77.46	$32.80	$59.83	$53.12	$48.74	$36.26	$52.47
Product Cost ($/equivalent bbl)										
Feed	$35.99	$16.23	$0.00	$23.64	$46.20	$19.15	$50.00	$21.65	$84.61	$8.00	$12.00	$0.00	$12.17
O&M total	$4.20	$11.78	$13.61	$17.21	$7.54	$15.81	$5.75	$7.72	$17.48	$14.52	$12.94	$14.05	$15.67
Capital charge	$13.85	$44.45	$50.08	$56.09	$28.46	$59.68	$21.71	$3.44	$29.97	$31.30	$23.80	$22.62	$25.79
By-products	$0.00	$0.00	$0.00	$0.00	$0.00	$0.00	$0.00	$0.00	($42.30)	($0.70)	$0.00	($0.41)	($1.16)
TOTAL	$54.05	$72.46	$63.69	$96.93	$82.20	$94.64	$77.46	$32.80	$89.75	$53.12	$48.74	$36.26	$52.47
Gasoline Equivalent Costs										
$/gal	$1.287	$1.725	$1.516	$2.308	$1.957	$2.253	$1.844	$0.781	$2.137	$1.265	$1.161	$0.863	$1.249
$/bbl	$54.048	$72.459	$63.693	$96.927	$82.196	$94.641	$77.464	$32.802	$89.750	$53.123	$48.741	$36.261	$52.466
Distribution/Marketing ($/bbl)	$7.567	$7.878	$7.730	$8.293	$4.799	$5.009	$4.719	($8.198)	$6.960				
Refining Credit	($10.848)	($13.681)	($12.332)	($17.445)	($14.261)	($16.176)	($13.533)	($4.132)	($15.731)			($1.400)	($5.900)
Gasoline Distribution Marketing Total	($4.423)	($4.735)	($4.587)	($5.149)	($4.799)	($5.009)	($4.719)	($12.33)	($4.960)				
TOTAL COST ($/bbl)	($7.70)	($10.54)	($9.19)	($14.30)	($14.26)	($16.18)	($13.53)	$20.47	($13.73)	$0.00	$0.00	($1.40)	($5.90)
	$46.34	$61.92	$54.50	$82.63	$67.94	$78.47	$63.93		$76.02	$53.12	$48.74	$34.86	$46.57
Vehicle Cost ($/gal additional)	$0.07	$0.07	$0.07	$0.07	$0.00	$0.00	$0.00	$0.34	$0.00	$0.00	$0.00	$0.00	$0.00
($/bbl additional)	$2.82	$2.82	$2.82	$2.82	$0.00	$0.00	$0.00	$14.12	$0.00	$0.00	$0.00	$0.00	$0.00
COST ($/oil bbl equivalent)	$49.17	$64.75	$57.33	$85.45	$67.94	$78.47	$63.93	$34.59	$76.02	$53.12	$48.74	$34.86	$46.57
Prices of Feedstocks										
Crude oil ($/bbl)	$49.17	$64.75	$57.33	$85.45	$67.94	$78.47	$63.93	$34.59	$76.02	$53.12	$48.74	$34.86	$46.57
Natural gas ($/Mcf)	$5.00	$5.00	$5.00	$5.00	$5.00	$5.00	$5.00	$4.30	$5.00	$5.00	$5.00	$4.31	$5.00
Corn ($/bu)	$2.88	$3.16	$3.02	$3.53	$3.21	$3.40	$3.14	$2.62	$3.36	$2.95	$2.87	$2.62	$2.83
Electricity ($/kWh)	$0.053	$0.056	$0.055	$0.061	$0.057	$0.059	$0.056	$0.050	$0.059	$0.054	$0.053	$0.050	$0.053

TABLE D-5 Cost Calculations for Conversion Technologies at 10 and 15 Percent Discount Rates ($20 and $40 per barrel, crude oil price)

			Feed/Fuel or Process										
Cost Factors	NG/ Methanol	Coal/ Methanol	Underground Coal Gasification/ Methanol	Wood/ Methanol	NG/ MTG	Coal/ MTG	NG/ Shell MDS	NG/ CNG	Corn/ Ethanol	Oil Shale/ Fluid Bed	Tar Sands/ Pyrolysis	Tar Sands/ Extraction	Coal/ Liquefaction
10 PERCENT DISCOUNT RATE ($40/BARREL)													
Product Cost ($/gal)													
Feed	0.439	$0.215	$0.000	$0.313	$1.015	$0.456	$1.098	$0.554	$1.086	$0.190	$0.286	$0.000	$0.288
O&M, nonenergy	0.056	$0.156	$0.180	$0.197	$0.180	$0.377	$0.137	$0.150	$0.156	$0.214	$0.167	$0.335	$0.247
O&M, energy	0.000	$0.000	$0.000	$0.026	$0.000	$0.000	$0.000	$0.034	$0.114	$0.123	$0.132	$0.000	$0.123
Capital charge	0.124	$0.398	$0.449	$0.502	$0.459	$0.962	$0.350	$0.055	$0.322	$0.505	$0.384	$0.365	$0.416
By-products	$0.000	$0.000	$0.000	$0.000	$0.000	$0.000	$0.000	$0.000	($0.543)	($0.017)	$0.000	($0.010)	($0.028)
TOTAL	$0.62	$0.77	$0.63	$1.04	$1.65	$1.79	$1.59	$0.79	$1.14	$1.02	$0.97	$0.69	$1.05
Product Cost ($/bbl)													
Feed	$18.45	$9.02	$0.00	$13.13	$42.62	$19.15	$46.13	$23.25	$45.60	$8.00	$12.00	$0.00	$12.09
O&M, nonenergy	$2.34	$6.54	$7.56	$8.26	$7.54	$15.81	$5.75	$6.30	$6.55	$9.00	$7.00	$14.05	$10.36
O&M, energy	$0.00	$0.00	$0.00	$1.10	$0.00	$0.00	$0.00	$1.45	$4.79	$5.16	$5.55	$0.00	$5.18
Capital charge	$5.21	$16.72	$18.84	$21.10	$19.27	$40.41	$14.70	$2.33	$13.53	$21.20	$16.12	$15.32	$17.46
By-products	$0.00	$0.00	$0.00	$0.00	$0.00	$0.00	$0.00	$0.00	($22.80)	($0.70)	$0.00	($0.41)	($1.16)
TOTAL	$25.99	$32.28	$26.40	$43.59	$69.43	$75.37	$66.58	$33.32	$47.67	$42.66	$40.67	$28.96	$43.93
Product Cost ($/equivalent bbl)													
Feed	$33.21	$16.23	$0.00	$23.64	$42.62	$19.15	$46.13	$23.25	$68.40	$8.00	$12.00	$0.00	$12.09
O&M total	$4.20	$11.78	$13.61	$16.85	$7.54	$15.81	$5.75	$7.75	$17.02	$14.16	$12.55	$14.05	$15.54
Capital charge	$9.38	$30.10	$33.91	$37.98	$19.27	$40.41	$14.70	$2.33	$20.29	$21.20	$16.12	$15.32	$17.46
By-products	$0.00	$0.00	$0.00	$0.00	$0.00	$0.00	$0.00	$0.00	($34.20)	($0.70)	$0.00	($0.41)	($1.16)
TOTAL	$46.79	$58.11	$47.52	$78.46	$69.43	$75.37	$66.58	$33.32	$71.51	$42.66	$40.67	$28.96	$43.93
Gasoline Equivalent Costs													
$/gal	$1.114	$1.383	$1.131	$1.868	$1.653	$1.795	$1.585	$0.793	$1.703	$1.016	$0.968	$0.689	$1.046
$/bbl	$46.789	$58.107	$47.520	$78.460	$69.431	$75.372	$66.582	$33.321	$71.509	$42.659	$40.669	$28.957	$43.929

Distribution/Marketing ($/bbl)	$7.456	$7.687	$7.471	$8.103	$4.653	$4.774	$4.595	$6.736			
Refining Credit	($9.182)	($9.182)	($9.182)	($9.182)	($9.182)	($9.182)	($9.182)	($9.182)			
Gasoline Distribution/Marketing	($4.240)	($4.240)	($4.240)	($4.240)	($4.240)	($4.240)	($4.240)	($4.240)		($1.400)	($5.900)
Total	($5.97)	($5.73)	($5.95)	($5.32)	($8.77)	($8.65)	($8.83)	($6.69)	$0.00	($1.40)	($5.90)
TOTAL COST ($/bbl)	$40.82	$52.37	$41.57	$73.14	$60.66	$66.72	$57.76	$64.82	$42.66	$27.56	$38.03
Vehicle Cost ($/gal additional)	$0.07	$0.07	$0.07	$0.07	$0.00	$0.00	$0.00	$0.00	$0.00	$0.00	$0.00
($/bbl additional)	$2.82	$2.82	$2.82	$2.82	$0.00	$0.00	$0.00	$0.00	$0.00	$0.00	$0.00
COST ($/oil bbl equivalent)	$43.65	$55.20	$44.39	$75.96	$60.66	$66.72	$57.76	$64.82	$42.66	$27.56	$38.03
Prices of Feedstocks											
Crude oil ($/bbl)	$40.00	$40.00	$40.00	$40.00	$40.00	$40.00	$40.00	$40.00	$40.00	$40.00	$40.00
Natural gas ($/Mcf)	$4.61	$4.61	$4.61	$4.61	$4.61	$4.61	$4.61	$4.61	$4.61	$4.61	$4.61
Corn ($/bu)	$2.71	$2.71	$2.71	$2.71	$2.71	$2.71	$2.71	$2.71	$2.71	$2.71	$2.71
Electricity ($/kWh)	$0.051	$0.051	$0.051	$0.051	$0.051	$0.051	$0.051	$0.051	$0.051	$0.051	$0.051

15 PERCENT DISCOUNT RATE ($40/BARREL)

Product Cost ($/gal)											
Feed	$0.439	$0.215	$0.000	$0.313	$1.015	$0.456	$1.098	$1.086	$0.190	$0.000	$0.288
O&M, nonenergy	$0.056	$0.156	$0.180	$0.197	$0.180	$0.377	$0.137	$0.156	$0.214	$0.335	$0.247
O&M, energy	$0.000	$0.000	$0.000	$0.026	$0.000	$0.000	$0.000	$0.114	$0.123	$0.000	$0.123
Capital charge	$0.183	$0.588	$0.662	$0.742	$0.678	$1.421	$0.517	$0.476	$0.567	$0.539	$0.614
By-products	$0.000	$0.000	$0.000	$0.000	$0.000	$0.000	$0.000	($0.543)	($0.017)	($0.010)	($0.028)
TOTAL	$0.68	$0.96	$0.84	$1.28	$1.87	$2.25	$1.75	$1.29	$1.26	$0.86	$1.24
Product Cost ($/bbl)											
Feed	$18.45	$9.02	$0.00	$13.13	$42.62	$19.15	$46.13	$45.60	$8.00	$0.00	$12.09
O&M, nonenergy	$2.34	$6.54	$7.56	$8.26	$7.54	$15.81	$5.75	$6.55	$9.00	$14.05	$10.36
O&M, energy	$0.00	$0.00	$0.00	$1.10	$0.00	$0.00	$0.00	$4.79	$5.16	$0.00	$5.18
Capital charge	$7.69	$24.69	$27.82	$31.16	$28.46	$59.68	$21.71	$19.98	$31.30	$22.62	$25.79
By-products	$0.00	$0.00	$0.00	$0.00	$0.00	$0.00	$0.00	($22.80)	($0.70)	($0.41)	($1.16)
TOTAL	$28.48	$40.26	$35.38	$53.65	$78.62	$94.64	$73.59	$54.12	$52.77	$36.26	$52.26

TABLE D-5 Continues

TABLE D-5 Continued

							Feed/Fuel or Process					
Cost Factors	NG/ Methanol	Coal/ Methanol	Underground Coal Gasi- fication/ Methanol	Wood/ Methanol	NG/ MTG	Coal/ MTG	NG/ Shell MDS	NG/ CNG	Corn/ Ethanol	Oil Shale/ Fluid Bed	Tar Sands/ Pyrolysis	Tar Sands/ Extraction	Coal/ Lique- faction
Product Cost ($/equivalent bbl)													
Feed	$33.21	$16.23	$0.00	$23.64	$42.62	$19.15	$46.13	$23.25	$68.40	$8.00	$12.00	$0.00	$12.09
O&M total	$4.20	$11.78	$13.61	$16.85	$7.54	$15.81	$5.75	$7.75	$17.02	$14.16	$12.55	$14.05	$15.54
Capital charge	$13.85	$44.45	$50.08	$56.09	$28.46	$59.68	$21.71	$3.44	$29.97	$31.30	$23.80	$22.62	$25.79
By-products	$0.00	$0.00	$0.00	$0.00	$0.00	$0.00	$0.00	$0.00	($34.20)	($0.70)	$0.00	($0.41)	($1.16)
TOTAL	$51.26	$72.46	$63.69	$96.57	$78.62	$94.64	$73.59	$34.43	$81.19	$52.77	$48.35	$36.26	$52.26
Gasoline Equivalent Costs													
$/gal	$1.221	$1.725	$1.516	$2.299	$1.872	$2.253	$1.752	$0.820	$1.933	$1.256	$1.151	$0.863	$1.244
$/bbl	$51.261	$72.459	$63.693	$96.571	$78.619	$94.641	$73.593	$34.430	$81.185	$52.767	$48.354	$36.261	$52.255
Distribution/Marketing ($/bbl)	$7.548	$7.980	$7.801	$8.472	$4.841	$5.168	$4.738		$6.934				
Refining Credit	($9.182)	($9.182)	($9.182)	($9.182)	($9.182)	($9.182)	($9.182)	($9.182)	($9.182)			($1.400)	($5.900)
Gasoline Distribution/Marketing Total	($4.240)	($4.240)	($4.240)	($4.240)	($4.240)	($4.240)	($4.240)	($4.240)	($4.240)				
	($5.87)	($5.44)	($5.62)	($4.95)	($8.58)	($8.25)	($8.68)	($13.42)	($6.49)	$0.00	$0.00	($1.40)	($5.90)
TOTAL COST ($/bbl)	$45.39	$67.02	$58.07	$91.62	$70.04	$86.39	$64.91	$21.01	$74.70	$52.77	$48.35	$34.86	$46.36
Vehicle Cost ($/gal additional)	$0.07	$0.07	$0.07	$0.07	$0.00	$0.00	$0.00	$0.34	$0.00	$0.00	$0.00	$0.00	$0.00
($/bbl additional)	$2.82	$2.82	$2.82	$2.82	$0.00	$0.00	$0.00	$14.12	$0.00	$0.00	$0.00	$0.00	$0.00
COST ($/bbl oil equivalent)	$48.21	$69.84	$60.90	$94.44	$70.04	$86.39	$64.91	$35.13	$74.70	$52.77	$48.35	$34.86	$46.36
Prices of Feedstocks													
Crude oil ($/bbl)	$40.00	$40.00	$40.00	$40.00	$40.00	$40.00	$40.00	$40.00	$40.00	$40.00	$40.00	$40.00	$40.00
Natural gas ($/Mcf)	$4.61	$4.61	$4.61	$4.61	$4.61	$4.61	$4.61	$4.61	$4.61	$4.61	$4.61	$4.61	$4.61
Corn ($/bu)	$2.71	$2.71	$2.71	$2.71	$2.71	$2.71	$2.71	$2.71	$2.71	$2.71	$2.71	$2.71	$2.71
Electricity ($/kWh)	$0.051	$0.051	$0.051	$0.051	$0.051	$0.051	$0.051	$0.051	$0.051	$0.051	$0.051	$0.051	$0.051

10 PERCENT DISCOUNT RATE ($20/BARREL)

Product Cost ($/gal)													
Feed	$0.328	$0.215	$0.000	$0.313	$0.757	$0.456	$0.819	$0.413	$0.943	$0.190	$0.286	$0.000	$0.282
O&M, nonenergy	$0.056	$0.156	$0.180	$0.197	$0.180	$0.377	$0.137	$0.150	$0.156	$0.214	$0.167	$0.335	$0.247
O&M, energy	$0.000	$0.000	$0.000	$0.024	$0.000	$0.000	$0.000	$0.032	$0.110	$0.098	$0.104	$0.000	$0.114
Capital charge	$0.124	$0.398	$0.449	$0.502	$0.459	$0.962	$0.350	$0.055	$0.322	$0.505	$0.384	$0.365	$0.416
By-products	$0.000	$0.000	$0.000	$0.000	$0.000	$0.000	$0.000	$0.000	($0.471)	($0.017)	$0.000	($0.010)	($0.028)
TOTAL	$0.51	$0.77	$0.63	$1.04	$1.40	$1.79	$1.31	$0.65	$1.06	$0.99	$0.94	$0.69	$1.03
Product Cost ($/bbl)													
Feed	$13.76	$9.02	$0.00	$13.13	$31.80	$19.15	$34.41	$17.34	$39.60	$8.00	$12.00	$0.00	$11.86
O&M, nonenergy	$2.34	$6.54	$7.56	$8.26	$7.54	$15.81	$5.75	$6.30	$6.55	$9.00	$7.00	$14.05	$10.36
O&M, energy	$0.00	$0.00	$0.00	$1.02	$0.00	$0.00	$0.00	$1.33	$4.62	$4.10	$4.38	$0.00	$4.77
Capital charge	$5.21	$16.72	$18.84	$21.10	$19.27	$40.41	$14.70	$2.33	$13.53	$21.20	$16.12	$15.32	$17.46
By-products	$0.00	$0.00	$0.00	$0.00	$0.00	$0.00	$0.00	$0.00	($19.80)	($0.70)	$0.00	($0.41)	($1.16)
TOTAL	$21.31	$32.28	$26.40	$43.50	$58.61	$75.37	$54.87	$27.30	$44.50	$41.60	$39.50	$28.96	$43.29
Product Cost ($/equivalent bbl)													
Feed	$24.77	$16.23	$0.00	$23.64	$31.80	$19.15	$34.41	$17.34	$59.40	$8.00	$12.00	$0.00	$11.86
O&M total	$4.20	$11.78	$13.61	$16.69	$7.54	$15.81	$5.75	$7.63	$16.76	$13.10	$11.38	$14.05	$15.13
Capital charge	$9.38	$30.10	$33.91	$37.98	$19.27	$40.41	$14.70	$2.33	$20.29	$21.20	$16.12	$15.32	$17.46
By-products	$0.00	$0.00	$0.00	$0.00	$0.00	$0.00	$0.00	$0.00	($29.70)	($0.70)	$0.00	($0.41)	($1.16)
TOTAL	$38.36	$58.11	$47.52	$78.30	$58.61	$75.37	$54.87	$27.30	$66.75	$41.60	$39.50	$28.96	$43.29
Gasoline Equivalent Costs													
$/gal	$0.913	$1.383	$1.131	$1.864	$1.395	$1.795	$1.306	$0.650	$1.589	$0.990	$0.940	$0.689	$1.031
$/bbl	$38.356	$58.107	$47.520	$78.303	$58.607	$75.372	$54.868	$27.303	$66.754	$41.599	$39.497	$28.957	$43.286
Distribution/Marketing ($/bbl)	$7.367	$7.770	$7.554	$8.182	$4.515	$4.857	$4.438		$6.722				
Refining Credit	($5.545)	($5.545)	($5.545)	($5.545)	($5.545)	($5.545)	($5.545)	($5.545)	($5.545)			($1.400)	($5.900)
Gasoline Distribution/Marketing	($3.840)	($3.840)	($3.840)	($3.840)	($3.840)	($3.840)	($3.840)	($3.840)	($3.840)				
Total	($2.02)	($1.62)	($1.83)	($1.20)	($4.87)	($4.53)	($4.95)	($9.39)	($2.66)	$0.00	$0.00	($1.40)	($5.90)
TOTAL COST ($/bbl)	$36.34	$56.49	$45.69	$77.10	$53.74	$70.84	$49.92	$17.92	$64.09	$41.60	$39.50	$27.56	$37.39

TABLE D-5 Continues

TABLE D-5 Continued

Cost Factors	NG/ Methanol	Coal/ Methanol	Underground Coal Gasification/ Methanol	Wood/ Methanol	NG/ MTG	Coal/ MTG	NG/ Shell MDS	NG/ CNG	Corn/ Ethanol	Oil Shale/ Fluid Bed	Tar Sands/ Pyrolysis	Tar Sands/ Extraction	Coal/ Liquefaction
Vehicle Cost ($/gal additional)	$0.07	$0.07	$0.07	$0.07	$0.00	$0.00	$0.00	$0.34	$0.00	$0.00	$0.00	$0.00	$0.00
($/bbl additional)	$2.82	$2.82	$2.82	$2.82	$0.00	$0.00	$0.00	$14.12	$0.00	$0.00	$0.00	$0.00	$0.00
COST ($/oil bbl equivalent)	$39.16	$59.31	$48.51	$79.92	$53.74	$70.84	$49.92	$32.03	$64.09	$41.60	$39.50	$27.56	$37.39
Prices of Feedstocks													
Crude oil ($/bbl)	$20.00	$20.00	$20.00	$20.00	$20.00	$20.00	$20.00	$20.00	$20.00	$20.00	$20.00	$20.00	$20.00
Natural gas ($/Mcf)	$3.44	$3.44	$3.44	$3.44	$3.44	$3.44	$3.44	$3.44	$3.44	$3.44	$3.44	$3.44	$3.44
Corn ($/bu)	$2.36	$2.36	$2.36	$2.36	$2.36	$2.36	$2.36	$2.36	$2.36	$2.36	$2.36	$2.36	$2.36
Electricity ($/kWh)	$0.047	$0.047	$0.047	$0.047	$0.047	$0.047	$0.047	$0.047	$0.047	$0.047	$0.047	$0.047	$0.047
15 PERCENT DISCOUNT RATE ($20/BARREL)													
Product Cost ($/gal)													
Feed	$0.328	$0.215	$0.000	$0.313	$0.757	$0.456	$0.819	$0.413	$0.943	$0.190	$0.286	$0.000	$0.282
O&M, nonenergy	$0.056	$0.156	$0.180	$0.197	$0.180	$0.377	$0.137	$0.150	$0.156	$0.214	$0.167	$0.335	$0.247
O&M, energy	$0.000	$0.000	$0.000	$0.024	$0.000	$0.000	$0.000	$0.032	$0.110	$0.098	$0.104	$0.000	$0.114
Capital charge	$0.183	$0.588	$0.662	$0.742	$0.678	$1.421	$0.517	$0.082	$0.476	$0.745	$0.567	$0.539	$0.614
By-products	$0.000	$0.000	$0.000	$0.000	$0.000	$0.000	$0.000	$0.000	($0.471)	($0.017)	$0.000	($0.010)	($0.028)
TOTAL	$0.57	$0.96	$0.84	$1.28	$1.61	$2.25	$1.47	$0.68	$1.21	$1.23	$1.12	$0.86	$1.23
Product Cost ($/bbl)													
Feed	$13.76	$9.02	$0.00	$13.13	$31.80	$19.15	$34.41	$17.34	$39.60	$8.00	$12.00	$0.00	$11.86
O&M, nonenergy	$2.34	$6.54	$7.56	$8.26	$7.54	$15.81	$5.75	$6.30	$6.55	$9.00	$7.00	$14.05	$10.36
O&M, energy	$0.00	$0.00	$0.00	$1.02	$0.00	$0.00	$0.00	$1.33	$4.62	$4.10	$4.38	$0.00	$4.77
Capital charge	$7.69	$24.69	$27.82	$31.16	$28.46	$59.68	$21.71	$3.44	$19.98	$31.30	$23.80	$22.62	$25.79
By-products	$0.00	$0.00	$0.00	$0.00	$0.00	$0.00	$0.00	$0.00	($19.80)	($0.70)	$0.00	($0.41)	($1.16)
TOTAL	$23.79	$40.26	$35.38	$53.56	$67.79	$94.64	$61.88	$28.41	$50.95	$51.71	$47.18	$36.26	$51.61

Product Cost ($/equivalent bbl)											
Feed	$16.23	$0.00	$23.64	$31.80	$19.15	$34.41	$17.34	$59.40	$8.00	$12.00	$0.00	$11.86	
O&M total	$11.78	$13.61	$16.69	$7.54	$15.81	$5.75	$7.63	$16.76	$13.10	$11.38	$14.05	$15.13	
Capital charge	13.85	$44.45	$50.08	$56.09	$28.46	$59.68	$21.71	$3.44	$29.97	$31.30	$23.80	$22.62	$25.79
By-products	$0.00	$0.00	$0.00	$0.00	$0.00	$0.00	$0.00	$0.00	($29.70)	($0.70)	$0.00	($0.41)	($1.16)
TOTAL	$42.83	$72.46	$63.69	$96.41	$67.79	$94.64	$61.88	$28.41	$76.43	$51.71	$47.18	$36.26	$51.61
Gasoline Equivalent Costs												
$/gal	$1.020	$1.725	$1.516	$2.296	$1.614	$2.253	$1.473	$0.676	$1.820	$1.231	$1.123	$0.863	$1.229
$/bbl	$42.828	$72.459	$63.693	$96.414	$67.795	$94.641	$61.878	$28.412	$76.430	$51.707	$47.183	$36.261	$51.612
Distribution/Marketing ($/bbl)	$7.458	$8.063	$7.884	$8.552	$4.702	$5.250	$4.581		$6.919				
Refining Credit	($5.545)	($5.545)	($5.545)	($5.545)	($5.545)	($5.545)	($5.545)	($5.545)	($5.545)			($1.400)	($5.900)
Gasoline Distribution/Marketing	($3.840)	($3.840)	($3.840)	($3.840)	($3.840)	($3.840)	($3.840)	($3.840)	($3.840)				
Total	($1.93)	($1.32)	($1.50)	($0.83)	($4.68)	($4.14)	($4.80)	($9.39)	($2.47)	$0.00	$0.00	($1.40)	($5.90)
TOTAL COST ($/bbl)	$40.90	$71.14	$62.19	$95.58	$63.11	$90.51	$57.07	$19.03	$73.96	$51.71	$47.18	$34.86	$45.71
Vehicle Cost ($/gal additional)	$0.07	$0.07	$0.07	$0.07	$0.00	$0.00	$0.00	$0.34	$0.00	$0.00	$0.00	$0.00	$0.00
($/bbl additional)	$2.82	$2.82	$2.82	$2.82	$0.00	$0.00	$0.00	$14.12	$0.00	$0.00	$0.00	$0.00	$0.00
COST ($/oil bbl equivalent)	$43.72	$73.96	$65.01	$98.40	$63.11	$90.51	$57.07	$33.14	$73.96	$51.71	$47.18	$34.86	$45.71
Prices of Feedstocks												
Crude oil ($/bbl)	$20.00	$20.00	$20.00	$20.00	$20.00	$20.00	$20.00	$20.00	$20.00	$20.00	$20.00	$20.00	$20.00
Natural gas ($/Mcf)	$3.44	$3.44	$3.44	$3.44	$3.44	$3.44	$3.44	$3.44	$3.44	$3.44	$3.44	$3.44	$3.44
Corn ($/bu)	$2.36	$2.36	$2.36	$2.36	$2.36	$2.36	$2.36	$2.36	$2.36	$2.36	$2.36	$2.36	$2.36
Electricity ($/kWh)	$0.047	$0.047	$0.047	$0.047	$0.047	$0.047	$0.047	$0.047	$0.047	$0.047	$0.047	$0.047	$0.047

TABLE D-6 Cost Calculations for Natural Gas Price $2/MMBtu Lower at 10 Percent Discount Rate (endogenous), 10 Percent Discount Rate at $40/Barrel Oil Price (exogenous), and 10 Percent Discount Rate at $20/Barrel Oil Price (exogenous)

			Feed/Fuel or Process										
Cost Factors	NG/ Methanol	Coal/ Methanol	Underground Coal Gasification/ Methanol	Wood/ Methanol	NG/ MTG	Coal/ MTG	NG/ Shell MDS	NG/ CNG	Corn/ Ethanol	Oil Shale/ Fluid Bed	Tar Sands/ Pyrolysis	Tar Sands/ Extraction	Coal/ Liquefaction
10 PERCENT DISCOUNT RATE (ENDOGENOUS)													
Product Cost ($/gal)													
Feed	$0.169	$0.215	$0.000	$0.313	$0.610	$0.456	$0.561	$0.188	$1.270	$0.190	$0.286	$0.000	$0.278
O&M, nonenergy	$0.056	$0.156	$0.180	$0.197	$0.180	$0.377	$0.137	$0.150	$0.156	$0.214	$0.167	$0.335	$0.247
O&M, energy	$0.000	$0.000	$0.000	$0.029	$0.000	$0.000	$0.000	$0.032	$0.119	$0.082	$0.083	$0.000	$0.122
Capital charge	$0.124	$0.398	$0.449	$0.502	$0.459	$0.962	$0.350	$0.055	$0.322	$0.505	$0.384	$0.365	$0.416
By-products	$0.000	$0.000	$0.000	$0.000	$0.000	$0.000	$0.000	$0.000	($0.635)	($0.017)	$0.000	($0.010)	($0.028)
TOTAL	$0.35	$0.77	$0.63	$1.04	$1.25	$1.79	$1.05	$0.43	$1.23	$0.97	$0.92	$0.69	$1.03
Product Cost ($/bbl)													
Feed	$7.11	$9.02	$0.00	$13.13	$25.64	$19.15	$23.57	$7.91	$53.33	$8.00	$12.00	$0.00	$11.66
O&M, nonenergy	$2.34	$6.54	$7.56	$8.26	$7.54	$5.81	$5.75	$6.30	$6.55	$9.00	$7.00	$14.05	$10.36
O&M, energy	$0.00	$0.00	$0.00	$1.23	$0.00	$0.00	$0.00	$1.34	$5.01	$3.43	$3.47	$0.00	$5.13
Capital charge	$5.21	$16.72	$18.84	$21.10	$19.27	$40.41	$14.70	$2.33	$13.53	$21.20	$16.12	$15.32	$17.46
By-products	$0.00	$0.00	$0.00	$0.00	$0.00	$0.00	$0.00	$0.00	($26.67)	($0.70)	$0.00	($0.41)	($1.16)
TOTAL	$14.66	$32.28	$26.40	$43.72	$52.44	$75.37	$44.03	$17.88	$51.76	$40.93	$38.59	$28.96	$43.45
Product Cost ($/equivalent bbl)													
Feed	$12.81	$16.23	$0.00	$23.64	$25.64	$19.15	$23.57	$7.91	$80.00	$8.00	$12.00	$0.00	$11.66
O&M total	$4.20	$11.78	$13.61	$17.08	$7.54	$15.81	$5.75	$7.64	$17.35	$12.43	$10.47	$14.05	$15.49
Capital charge	$9.38	$30.10	$33.91	$37.98	$19.27	$40.41	$14.70	$2.33	$20.29	$21.20	$16.12	$15.32	$17.46
By-products	$0.00	$0.00	$0.00	$0.00	$0.00	$0.00	$0.00	$0.00	($40.00)	($0.70)	$0.00	($0.41)	($1.16)
TOTAL	$26.39	$58.11	$47.52	$78.70	$52.44	$75.37	$44.03	$17.88	$77.64	$40.93	$38.59	$28.96	$43.45
Gasoline Equivalent Costs													
$/gal	$0.628	$1.383	$1.131	$1.874	$1.249	$1.795	$1.048	$0.426	$1.848	$0.974	$0.919	$0.689	$1.035
$/bbl	$26.388	$58.107	$47.520	$78.696	$52.445	$75.372	$44.028	$17.875	$77.637	$40.928	$38.586	$28.957	$43.450

	C1	C2	C3	C4	C5	C6	C7	C8	C9	C10	C11	C12	C13
Distribution/Marketing ($/bbl)	$7.099	$7.636	$7.456	$7.984	$4.295	$4.683	$4.153		$6.755				
Refining Credit	($6.593)	($11.473)	($9.844)	($14.640)	($9.684)	($13.211)	($8.389)	($5.940)	($13.867)			($1.400)	($5.900)
Gasoline Distribution/Marketing Total	($3.955)	($4.492)	($4.313)	($4.840)	($4.295)	($4.683)	($4.153)	($3.883)	($4.755)				
Total	($3.45)	($8.33)	($6.70)	($11.50)	($9.68)	($13.21)	($8.39)	($9.82)	($11.87)	$0.00	$0.00	($1.40)	($5.90)
TOTAL COST ($/bbl)	$22.94	$49.78	$40.82	$67.20	$42.76	$62.16	$35.64	$8.05	$65.77	$40.93	$38.59	$27.56	$37.55
Vehicle Cost ($/gal additional)	$0.07	$0.07	$0.07	$0.07	$0.00	$0.00	$0.00	$0.34	$0.00	$0.00	$0.00	$0.00	$0.00
($/bbl additional)	$2.82	$2.82	$2.82	$2.82	$0.00	$0.00	$0.00	$14.12	$0.00	$0.00	$0.00	$0.00	$0.00
COST ($/oil bbl equivalent)	$25.76	$52.60	$43.64	$70.02	$42.76	$62.16	$35.64	$22.17	$65.77	$40.93	$38.59	$27.56	$37.55
Prices of Feedstocks													
Crude oil ($/bbl)	$25.76	$52.60	$43.64	$70.02	$42.76	$62.16	$35.64	$22.17	$65.77	$40.93	$38.59	$27.56	$37.55
Natural gas ($/Mcf)	$1.78	$3.00	$2.83	$3.00	$2.77	$3.00	$2.36	$1.57	$3.00	$2.67	$2.53	$1.88	$2.47
Corn ($/bu)	$2.46	$2.94	$2.78	$3.25	$2.76	$3.11	$2.64	$2.40	$3.17	$2.73	$2.69	$2.49	$2.67
Electricity ($/kWh)	$0.049	$0.054	$0.052	$0.057	$0.052	$0.056	$0.051	$0.048	$0.057	$0.052	$0.051	$0.049	$0.051

10 PERCENT DISCOUNT RATE ($40/BARREL EXOGENOUS)

Product Cost ($/gal)													
Feed	$0.249	$0.215	$0.000	$0.313	$0.575	$0.456	$0.622	$0.314	$1.086	$0.190	$0.286	$0.000	$0.278
O&M, nonenergy	$0.056	$0.156	$0.180	$0.197	$0.180	$0.377	$0.137	$0.150	$0.156	$0.214	$0.167	$0.335	$0.247
O&M, energy	$0.000	$0.000	$0.000	$0.026	$0.000	$0.000	$0.000	$0.034	$0.114	$0.081	$0.085	$0.000	$0.123
Capital charge	$0.124	$0.398	$0.449	$0.502	$0.459	$0.962	$0.350	$0.055	$0.322	$0.505	$0.384	$0.365	$0.416
By-products	$0.000	$0.000	$0.000	$0.000	$0.000	$0.000	$0.000	$0.000	($0.543)	($0.017)	$0.000	($0.010)	($0.028)
TOTAL	$0.43	$0.77	$0.63	$1.04	$1.21	$1.79	$1.11	$0.55	$1.14	$0.97	$0.92	$0.69	$1.04

Product Cost ($/bbl)													
Feed	$10.45	$9.02	$0.00	$13.13	$24.14	$19.15	$26.13	$13.17	$45.60	$8.00	$12.00	$0.00	$11.69
O&M, nonenergy	$2.34	$6.54	$7.56	$8.26	$7.54	$15.81	$5.75	$6.30	$6.55	$9.00	$7.00	$14.05	$10.36
O&M, energy	$0.00	$0.00	$0.00	$1.10	$0.00	$0.00	$0.00	$1.45	$4.79	$3.38	$3.55	$0.00	$5.18
Capital charge	$5.21	$16.72	$18.84	$21.10	$19.27	$40.41	$14.70	$2.33	$13.53	$21.20	$16.12	$15.32	$17.46
By-products	$0.00	$0.00	$0.00	$0.00	$0.00	$0.00	$0.00	$0.00	($22.80)	($0.70)	$0.00	($0.41)	($1.16)
TOTAL	$18.00	$32.28	$26.40	$43.59	$50.95	$75.37	$46.58	$23.24	$47.67	$40.88	$38.67	$28.96	$43.53

TABLE D-6 Continues

TABLE D-6 Continued

			Feed/Fuel or Process									
Cost Factors	NG/ Methanol	Coal/ Methanol	Underground Coal Gasi- fication/ Methanol	Wood/ Methanol	NG/ MTG	Coal/ MTG	NG/ Shell MDS	NG/ CNG	Corn/ Ethanol	Oil Shale/ Fluid Bed	Tar Sands/ Pyrolysis	Tar Sands/ Extraction	Coal/ Lique- faction
Product Cost ($/equivalent bbl)													
Feed	$18.81	$16.23	$0.00	$23.64	$24.14	$19.15	$26.13	$13.17	$68.40	$8.00	$12.00	$0.00	$11.69
O&M total	$4.20	$11.78	$13.61	$16.85	$7.54	$15.81	$5.75	$7.75	$17.02	$12.38	$10.55	$14.05	$15.54
Capital charge	$9.38	$30.10	$33.91	$37.98	$19.27	$40.41	$14.70	$2.33	$20.29	$21.20	$16.12	$15.32	$17.46
By-products	$0.00	$0.00	$0.00	$0.00	$0.00	$0.00	$0.00	$0.00	($34.20)	($0.70)	$0.00	($0.41)	($1.16)
TOTAL	$32.39	$58.11	$47.52	$78.46	$50.95	$75.37	$46.58	$23.24	$71.51	$40.88	$38.67	$28.96	$43.53
Gasoline Equivalent Costs													
$/gal	$0.771	$1.383	$1.131	$1.868	$1.213	$1.795	$1.109	$0.553	$1.703	$0.973	$0.921	$0.689	$1.036
$/bbl	$32.391	$58.107	$47.520	$78.460	$50.951	$75.372	$46.582	$23.241	$71.509	$40.879	$38.669	$28.957	$43.529
Distribution/Marketing ($/bbl)	$7.163	$7.687	$7.471	$8.103	$4.276	$4.774	$4.187		$6.736				
Refining credit	($9.182)	($9.182)	($9.182)	($9.182)	($9.182)	($9.182)	($9.182)	($9.182)	($9.182)			($1.400)	($5.900)
Gasoline distribution/marketing	($4.240)	($4.240)	($4.240)	($4.240)	($4.240)	($4.240)	($4.240)	($4.240)	($4.240)				
Total	($6.26)	($5.73)	($5.95)	($5.32)	($9.15)	($8.65)	($9.23)	($13.42)	($6.69)	$0.00	$0.00	($1.40)	($5.90)
TOTAL COST ($/bbl)	$26.13	$52.37	$41.57	$73.14	$41.80	$66.72	$37.35	$9.82	$64.82	$40.88	$38.67	$27.56	$37.63
Vehicle Cost ($/gal additional)	$0.07	$0.07	$0.07	$0.07	$0.00	$0.00	$0.00	$0.34	$0.00	$0.00	$0.00	$0.00	$0.00
($/bbl additional)	$2.82	$2.82	$2.82	$2.82	$0.00	$0.00	$0.00	$14.12	$0.00	$0.00	$0.00	$0.00	$0.00
COST ($/oil bbl equivalent)	$28.96	$55.20	$44.39	$75.96	$41.80	$66.72	$37.35	$23.94	$64.82	$40.88	$38.67	$27.56	$37.63
Prices of Feedstocks													
Crude oil ($/bbl)	$40.00	$40.00	$40.00	$40.00	$40.00	$40.00	$40.00	$40.00	$40.00	$40.00	$40.00	$40.00	$40.00
Natural gas ($/Mcf)	$2.61	$2.61	$2.61	$2.61	$2.61	$2.61	$2.61	$2.61	$2.61	$2.61	$2.61	$2.61	$2.61
Corn ($/bu)	$2.71	$2.71	$2.71	$2.71	$2.71	$2.71	$2.71	$2.71	$2.71	$2.71	$2.71	$2.71	$2.71
Electricity ($/kWh)	$0.051	$0.051	$0.051	$0.051	$0.051	$0.051	$0.051	$0.051	$0.051	$0.051	$0.051	$0.051	$0.051

10 PERCENT DISCOUNT RATE ($20/BARREL EXOGENOUS)

Product Cost ($/gal)													
Feed	$0.137	$0.215	$0.000	$0.313	$0.317	$0.456	$0.343	$0.173	$0.943	$0.190	$0.286	$0.000	$0.273
O&M, nonenergy	$0.056	$0.156	$0.180	$0.197	$0.180	$0.377	$0.137	$0.150	$0.156	$0.214	$0.167	$0.335	$0.247
O&M, energy	$0.000	$0.000	$0.000	$0.024	$0.000	$0.000	$0.000	$0.032	$0.110	$0.055	$0.057	$0.000	$0.114
Capital charge	$0.124	$0.398	$0.449	$0.502	$0.459	$0.962	$0.350	$0.055	$0.322	$0.505	$0.384	$0.365	$0.416
By-products	$0.000	$0.000	$0.000	$0.000	$0.000	$0.000	$0.000	$0.000	($0.471)	($0.017)	$0.000	($0.010)	($0.028)
TOTAL	$0.32	$0.77	$0.63	$1.04	$0.96	$1.79	$0.83	$0.41	$1.06	$0.95	$0.89	$0.69	$1.02
Product Cost ($/bbl)													
Feed	$5.76	$9.02	$0.00	$13.13	$13.32	$19.15	$14.41	$7.26	$39.60	$8.00	$12.00	$0.00	$11.46
O&M, nonenergy	$2.34	$6.54	$7.56	$8.26	$7.54	$15.81	$5.75	$6.30	$6.55	$9.00	$7.00	$14.05	$10.36
O&M, energy	$0.00	$0.00	$0.00	$1.02	$0.00	$0.00	$0.00	$1.33	$4.62	$2.32	$2.38	$0.00	$4.77
Capital charge	$5.21	$16.72	$18.84	$21.10	$19.27	$40.41	$14.70	$2.33	$13.53	$21.20	$16.12	$15.32	$17.46
By-products	$0.00	$0.00	$0.00	$0.00	$0.00	$0.00	$0.00	$0.00	($19.80)	($0.70)	$0.00	($0.41)	($1.16)
TOTAL	$13.31	$32.28	$26.40	$43.50	$40.13	$75.37	$34.87	$17.22	$44.50	$39.82	$37.50	$28.96	$42.89
Product Cost ($/equivalent bbl)													
Feed	$10.38	$16.23	$0.00	$23.64	$13.32	$19.15	$14.41	$7.26	$59.40	$8.00	$12.00	$0.00	$11.46
O&M total	$4.20	$11.78	$13.61	$16.69	$7.54	$15.81	$5.75	$7.63	$16.76	$11.32	$9.38	$14.05	$15.13
Capital charge	$9.38	$30.10	$33.91	$37.98	$19.27	$40.41	$14.70	$2.33	$20.29	$21.20	$16.12	$15.32	$17.46
By-products	$0.00	$0.00	$0.00	$0.00	$0.00	$0.00	$0.00	$0.00	($29.70)	($0.70)	$0.00	($0.41)	($1.16)
TOTAL	$23.96	$58.11	$47.52	$78.30	$40.13	$75.37	$34.87	$17.22	$66.75	$39.82	$37.50	$28.96	$42.89
Gasoline Equivalent Costs													
$/gal	$0.570	$1.383	$1.131	$1.864	$0.955	$1.795	$0.830	$0.410	$1.589	$0.948	$0.893	$0.689	$1.021
$/bbl	$23.958	$58.107	$47.520	$78.303	$40.127	$75.372	$34.868	$17.223	$66.754	$39.819	$37.497	$28.957	$42.886
Distribution/Marketing ($/bbl)													
Refining credit	$7.073	($5.545)	$7.554	$8.182	$4.138	$4.857	$4.030	($5.545)	$6.722				($5.900)
Gasoline distribution/marketing	($3.840)	($3.840)	($3.840)	($3.840)	($3.840)	($3.840)	($3.840)	($3.840)	($3.840)			($1.400)	
Total	($2.31)	($1.62)	($1.83)	($1.20)	($5.25)	($4.53)	($5.36)	($9.39)	($2.66)	$0.00	$0.00	($1.40)	($5.90)
TOTAL COST ($/bbl)	$21.65	$56.49	$45.69	$77.10	$34.88	$70.84	$29.51	$7.84	$64.09	$39.82	$37.50	$27.56	$36.99

TABLE D-6 Continues

TABLE D-6 Continued

Cost Factors	Feed/Fuel or Process											
	NG/ Methanol	Coal/ Methanol	Underground Coal Gasification/ Methanol	Wood/ Methanol	NG/ MTG	Coal/ MTG	NG/ Shell MDS	NG/ CNG	Corn/ Ethanol	Oil Shale/ Fluid Bed	Tar Sands/ Pyrolysis	Tar Sands/ Extraction	Coal/ Liquefaction
Vehicle Cost ($/gal additional)	$0.07	$0.07	$0.07	$0.07	$0.00	$0.00	$0.00	$0.34	$0.00	$0.00	$0.00	$0.00	$0.00
($/bbl additional)	$2.82	$2.82	$2.82	$2.82	$0.00	$0.00	$0.00	$14.12	$0.00	$0.00	$0.00	$0.00	$0.00
COST ($/oil bbl equivalent)	$24.47	$59.31	$48.51	$79.92	$34.88	$70.84	$29.51	$21.95	$64.09	$39.82	$37.50	$27.56	$36.99
Prices of Feedstocks													
Crude oil ($/bbl)	$20.00	$20.00	$20.00	$20.00	$20.00	$20.00	$20.00	$20.00	$20.00	$20.00	$20.00	$20.00	$20.00
Natural gas ($/Mcf)	$1.44	$1.44	$1.44	$1.44	$1.44	$1.44	$1.44	$1.44	$1.44	$1.44	$1.44	$1.44	$1.44
Corn ($/bu)	$2.36	$2.36	$2.36	$2.36	$2.36	$2.36	$2.36	$2.36	$2.36	$2.36	$2.36	$2.36	$2.36
Electricity ($/kWh)	$0.047	$0.047	$0.047	$0.047	$0.047	$0.047	$0.047	$0.047	$0.047	$0.047	$0.047	$0.047	$0.047

TABLE D-7 Sensitivity Studies—Natural Gas to Methanol in Foreign or Remote Locations (endogenous crude oil price)

Remote?	No	Yes	Yes	Yes	Yes	Yes	Yes	Yes	Yes	Yes	Yes	Yes	Yes	Yes	Yes
Capital Cost Multiplier	1.00	1.00	1.00	1.25	1.25	1.25	1.25	1.25	1.25	1.75	1.75	1.75	1.75	1.75	1.75
Cost of Capital	10.0%	10.0%	10.0%	10.0%	10.0%	10.0%	15.0%	15.0%	15.0%	10.0%	10.0%	10.0%	5.0%	5.0%	5.0%
Capital Charge Rate	16.2%	16.2%	16.2%	16.2%	16.2%	16.2%	23.9%	23.9%	23.9%	16.2%	16.2%	16.2%	10.0%	10.0%	10.0%
Gasoline Equivalency	1.80	1.80	1.80	1.80	1.80	1.80	1.80	1.80	1.80	1.80	1.80	1.80	1.80	1.80	1.80
Capacity (bbl/day oil equivalent)	43,907	43,907	43,907	43,907	43,907	43,907	43,907	43,907	43,907	43,907	43,907	43,907	43,907	43,907	43,907
Capacity (bbl/day actual)	79,033	79,033	79,033	79,033	79,033	79,033	79,033	79,033	79,033	79,033	79,033	79,033	79,033	79,033	79,033
Capacity Factor	95.0%	95.0%	95.0%	95.0%	95.0%	95.0%	95.0%	95.0%	95.0%	95.0%	95.0%	95.0%	95.0%	95.0%	95.0%
Average Production (bbl/day, actual)	75081	75081	75081	75081	75081	75081	75081	75081	75081	75081	75081	75081	75081	75081	75081
Capital ($MM)	$883	$883	$883	$1,104	$1,104	$1,104	$1,104	$1,104	$1,104	$1,545	$1,545	$1,545	$1,545	$1,545	$1,545
Capital/Capacity	$11,173	$11,173	$11,173	$13,966	$13,966	$13,966	$13,966	$13,966	$13,966	$19,552	$19,552	$19,552	$19,552	$19,552	$19,552
Capital/Capacity (oil equivalent)	$20,111	$20,111	$20,111	$25,138	$25,138	$25,138	$25,138	$25,138	$25,138	$35,194	$35,194	$35,194	$35,194	$35,194	$35,194
Vehicle Capital Cost	$200.00	$200.00	$200.00	$200.00	$200.00	$200.00	$200.00	$200.00	$200.00	$200.00	$200.00	$200.00	$200.00	$200.00	$200.00
Operating: Feedstocks															
Natural gas (Mcf/gal)	0.095	0.095	0.095	0.095	0.095	0.095	0.095	0.095	0.095	0.095	0.095	0.095	0.095	0.095	0.095
Wellhead (Mcf/bbl)	4.00	4.00	4.00	4.00	4.00	4.00	4.00	4.00	4.00	4.00	4.00	4.00	4.00	4.00	4.00
O&M															
Nonenergy ($/gal)	$0.06	$0.06	$0.06	$0.06	$0.06	$0.06	$0.06	$0.06	$0.06	$0.09	$0.09	$0.09	$0.09	$0.09	$0.09
Cost per barrel	$2.34	$2.34	$2.34	$2.48	$2.48	$2.48	$2.48	$2.48	$2.48	$3.83	$3.83	$3.83	$3.83	$3.83	$3.83
Product Cost ($/gal)															
Feed	$0.465	$0.286	$0.095	$0.286	$0.190	$0.095	$0.286	$0.190	$0.095	$0.286	$0.190	$0.095	$0.286	$0.190	$0.095
O&M	$0.056	$0.056	$0.056	$0.059	$0.059	$0.059	$0.059	$0.059	$0.059	$0.091	$0.091	$0.091	$0.091	$0.091	$0.091
Capital charge	$0.124	$0.124	$0.124	$0.155	$0.155	$0.155	$0.229	$0.229	$0.229	$0.217	$0.217	$0.217	$0.134	$0.134	$0.134
TOTAL	$0.65	$0.47	$0.27	$0.50	$0.40	$0.31	$0.57	$0.48	$0.38	$0.59	$0.50	$0.40	$0.51	$0.42	$0.32

TABLE D-7 Continues

TABLE D-7 Continued

Product Cost ($/bbl)										
Feed	$19.55	$12.00	$4.00	$12.00	$8.00	$12.00	$8.00	$4.00	$12.00	$4.00	$8.00	$4.00
O&M, nonenergy	$2.34	$2.34	$2.34	$2.48	$2.48	$2.48	$2.48	$2.48	$3.83	$3.83	$3.83	$3.83
Capital charge	$5.21	$5.21	$5.21	$6.51	$6.51	$9.62	$9.62	$9.62	$9.12	$9.12	$5.64	$5.64
TOTAL	$27.09	$19.54	$11.54	$20.99	$16.99	$24.10	$20.10	$16.10	$24.95	$16.95	$17.47	$13.47
Product Cost ($/equivalent bbl)											
Feed	$35.19	$21.60	$7.20	$21.60	$14.40	$21.60	$14.40	$7.20	$21.60	$7.20	$14.40	$7.20
O&M total	$4.20	$4.20	$4.20	$4.47	$4.47	$4.47	$4.47	$4.47	$6.90	$6.90	$6.90	$6.90
Capital charge	$9.38	$9.38	$9.38	$11.72	$11.72	$17.31	$17.31	$17.31	$16.41	$16.41	$10.14	$10.14
TOTAL	$48.77	$35.18	$20.78	$37.79	$30.59	$43.38	$36.18	$28.98	$44.90	$30.51	$31.44	$24.24
Gasoline Equivalent Costs											
$/gal	$1.161	$0.838	$0.495	$0.900	$0.728	$1.033	$0.861	$0.690	$1.069	$0.726	$0.749	$0.577
$/bbl	$48.770	$35.178	$20.781	$37.785	$30.587	$43.376	$36.177	$28.979	$44.905	$30.507	$31.439	$24.240
Distribution/Marketing ($/bbl)											
Refining Credit	$7.478	$7.234	$6.990	$7.278	$7.156	$7.373	$7.251	$7.129	$7.398	$7.277	$7.292	$7.171	$7.049
Refining Credit	($10.036)	($8.548)	($6.333)	($8.950)	($7.842)	($9.810)	($8.702)	($7.595)	($10.045)	($8.937)	($9.081)	($7.973)	($6.866)
Gasoline Distribution/Marketing	($4.334)	($4.170)	($3.927)	($4.214)	($4.093)	($4.309)	($4.187)	($4.065)	($4.335)	($4.213)	($4.229)	($4.107)	($3.985)
Total	($6.89)	($5.48)	($3.27)	($5.89)	($4.78)	($6.75)	($5.64)	($4.53)	($6.98)	($5.87)	($6.02)	($4.91)	($3.80)
TOTAL	$41.88	$29.69	$17.51	$31.90	$25.81	$36.63	$30.54	$24.45	$37.92	$31.83	$32.62	$26.53	$20.44
Vehicle Cost ($/gal additional)											
($/bbl additional)	$0.07	$0.07	$0.07	$0.07	$0.07	$0.07	$0.07	$0.07	$0.07	$0.07	$0.07	$0.07
Transportation Costs	$2.82	$2.82	$2.82	$2.82	$2.82	$2.82	$2.82	$2.82	$2.82	$2.82	$2.82	$2.82
COST ($/ oil bbl equivalent)	$0.00	$4.00	$4.00	$4.00	$4.00	$4.00	$4.00	$4.00	$4.00	$4.00	$4.00	$4.00
	$44.70	$36.52	$24.33	$38.72	$32.63	$43.45	$37.36	$31.27	$44.75	$38.66	$39.44	$33.35	$27.26
Prices of Feedstocks											
Crude oil ($/bbl)	$44.70	$36.52	$24.33	$38.72	$32.63	$43.45	$37.36	$31.27	$44.75	$38.66	$39.44	$33.35	$27.26
Natural gas ($/Mcf)	$4.89	$3.00	$1.00	$3.00	$2.00	$3.00	$2.00	$1.00	$3.00	$1.00	$3.00	$2.00	$1.00
Corn ($/bu)	$2.80	$2.65	$2.43	$2.69	$2.58	$2.78	$2.67	$2.56	$2.80	$2.58	$2.70	$2.60	$2.49
Electricity ($/kWh)	$0.052	$0.051	$0.048	$0.051	$0.050	$0.052	$0.051	$0.050	$0.052	$0.050	$0.050	$0.050	$0.049
Electricity (sold)	$0.042	$0.041	$0.038	$0.041	$0.040	$0.042	$0.041	$0.040	$0.042	$0.040	$0.041	$0.040	$0.039
Corn by-product ($/bu)	$1.39	$1.30	$1.16	$1.32	$1.25	$1.38	$1.31	$1.24	$1.39	$1.32	$1.33	$1.26	$1.19

TABLE D-8 Sensitivity Studies—Natural Gas to Methanol from Domestic Sources (endogenous crude oil price)

Cost Factors	Base Case	Efficiency Gain		Natural Gas Price		Investment Cost		Real Discount Rate		Differential Auto Costs	
		0%	15% More	Lower	Higher	Up 25%	Down 20%	5% DCF	15% DCF	$0	$500
Gasoline Equivalency	1.80	2.06	1.57	1.80	1.80	1.80	1.80	1.80	1.80	1.80	1.80
Capacity (bbl/day oil equivalent)	43,907	43,907	43,907	43,907	43,907	43,907	43,907	43,907	43,907	43,907	43,907
Capacity (bbl/day actual)	79,033	90,448	68,714	79,033	79,033	79,033	79,033	79,033	79,033	79,033	79,033
Capacity Factor	95.0%	95.0%	95.0%	95.0%	95.0%	95.0%	95.0%	95.0%	95.0%	95.0%	95.0%
Average Production (bbl/day actual)	75,081	85,926	65,279	75,081	75,081	75,081	75,081	75,081	75,081	75,081	75,081
Capital ($MM)	$883	$883	$883	$883	$883	$1,104	$706	$883	$883	$883	$883
Capital/Capacity ($MM/bbl/day actual)	$11,173	$9,762	$12,850	$11,173	$11,173	$13,966	$8,938	$11,173	$11,173	$11,173	$11,173
Capital/Capacity ($MM/bbl/day oil equivalent)	$20,111	$20,111	$20,111	$20,111	$20,111	$25,138	$16,089	$20,111	$20,111	$20,111	$20,111
Vehicle Capital Cost	$200.00	$200.00	$200.00	$200.00	$200.00	$200.00	$200.00	$200.00	$200.00	$0.00	$500.00
Operating: Feedstocks											
Natural gas (Mcf/gal)	0.095	0.095	0.095	0.095	0.095	0.095	0.095	0.095	0.095	0.095	0.095
Wellhead (Mcf/bbl)	4.00	4.00	4.00	4.00	4.00	4.00	4.00	4.00	4.00	4.00	4.00
O&M											
Nonenergy ($/gal)	$0.06	$0.06	$0.06	$0.06	$0.06	$0.06	$0.06	$0.06	$0.06	$0.06	$0.06
Cost per barrel	$2.34	$2.34	$2.34	$2.34	$2.34	$2.34	$2.34	$2.34	$2.34	$2.34	$2.34
Product Cost ($/gal)											
Feed	$0.465	$0.476	$0.424	$0.275	$0.656	$0.476	$0.452	$0.439	$0.476	$0.445	$0.476
O&M	$0.056	$0.056	$0.056	$0.056	$0.056	$0.056	$0.056	$0.056	$0.056	$0.056	$0.056
Capital charge	$0.124	$0.108	$0.143	$0.124	$0.124	$0.155	$0.099	$0.077	$0.183	$0.124	$0.124
TOTAL	$0.65	$0.64	$0.62	$0.45	$0.84	$0.69	$0.61	$0.57	$0.72	$0.62	$0.66

TABLE D-8 Continues

TABLE D-8 Continued

Cost Factors	Base Case	Efficiency Gain 0%	Efficiency Gain 15% More	Natural Gas Price Lower	Natural Gas Price Higher	Investment Cost Up 25%	Investment Cost Down 20%	Real Discount Rate 5% DCF	Real Discount Rate 15% DCF	Differential Auto Costs $0	Differential Auto Costs $500
Product Cost ($/bbl)											
Feed	$19.55	$20.00	$17.80	$11.55	$27.55	$20.00	$18.97	$18.45	$20.00	$18.68	$20.00
O&M	$2.34	$2.34	$2.34	$2.34	$2.34	$2.34	$2.34	$2.34	$2.34	$2.34	$2.34
Capital charge	$5.21	$4.55	$5.99	$5.21	$5.21	$6.51	$4.17	$3.22	$7.70	$5.21	$5.21
TOTAL	$27.10	$26.88	$26.13	$19.10	$35.09	$28.84	$25.48	$24.00	$30.03	$26.23	$27.54
Product Cost ($/equivalent bbl)											
Feed	$35.19	$41.19	$27.86	$20.79	$49.59	$35.99	$34.15	$33.20	$35.99	$33.63	$35.99
O&M total	$4.20	$4.81	$3.65	$4.20	$4.20	$4.20	$4.20	$4.20	$4.20	$4.20	$4.20
Capital charge	$9.38	$9.38	$9.38	$9.38	$9.38	$11.72	$7.50	$5.80	$13.86	$9.38	$9.38
TOTAL	$48.77	$55.38	$40.89	$34.37	$63.17	$51.92	$45.86	$43.21	$54.06	$47.21	$49.58
Gasoline Equivalent Costs											
$/gal	$1.161	$1.319	$0.974	$0.818	$1.504	$1.236	$1.092	$1.029	$1.287	$1.124	$1.180
$/bbl	$48.771	$55.382	$40.893	$34.374	$63.169	$51.920	$45.855	$43.208	$54.059	$47.214	$49.576
Distribution/Marketing ($/bbl)	$7.478	$8.647	$6.388	$7.234	$7.721	$7.531	$7.428	$7.383	$7.567	$7.462	$7.477
Refining Credit	($10.037)	($11.214)	($8.680)	($7.822)	($12.252)	($10.521)	($9.588)	($9.181)	($10.850)	($9.365)	($10.799)
Gasoline Distribution/Marketing Total	($4.334)	($4.464)	($4.185)	($4.090)	($4.578)	($4.387)	($4.285)	($4.240)	($4.424)	($4.260)	($4.418)
	($6.89)	($7.03)	($6.48)	($4.68)	($9.11)	($7.38)	($6.44)	($6.04)	($7.71)	($6.16)	($7.74)
TOTAL COST ($/bbl)	$41.88	$48.35	$34.42	$29.70	$54.06	$44.54	$39.41	$37.17	$46.35	$41.12	$41.84
Vehicle Cost ($/gal additional)	$0.07	$0.07	$0.07	$0.07	$0.07	$0.07	$0.07	$0.07	$0.07	$0.00	$0.17
($/bbl additional)	$2.82	$2.82	$2.82	$2.82	$2.82	$2.82	$2.82	$2.82	$2.82	$0.00	$7.06
COST ($/oil bbl equivalent)	$44.70	$51.18	$37.24	$32.52	$56.88	$47.37	$42.23	$39.99	$49.18	$41.12	$48.89
Prices of Feedstocks											
Crude oil ($/bbl)	$44.70	$51.18	$37.24	$32.52	$56.88	$47.37	$42.23	$39.99	$49.18	$41.01	$48.89
Natural gas ($/Mcf)	$4.89	$5.00	$4.45	$2.89	$6.89	$5.00	$4.74	$4.61	$5.00	$4.67	$5.00
Electricity price ($/kWh)	$0.052	$0.054	$0.051	$0.050	$0.055	$0.053	$0.052	$0.051	$0.053	$0.052	$0.053

TABLE D-9 Technologies and Source Data Used in the Economic Analysis

1. Natural gas to methanol: California Fuel Methanol Study (1989).
2. Coal to methanol: Data from Schulman and Biasca (1989) were adjusted by the committee to larger plant sizes of 50,000 bbl/day of oil equivalent using a scaling component of 0.85. This size plant is consistent with the scale used for natural gas to methanol, oil shale conversion, and direct coal liquefaction.
3. Underground coal gasification: Schulman and Biasca (1989).
4. Wood to methanol: Data from Reed and Graboski (1989). Methanol production from biomass and municipal waste: Process Design and Economics, Syn-Gas, Inc., Golden, Colorado.
5. Natural gas to gasoline through Mobil's fluid bed MTG process: Data of Schulman and Biasca (1989) adjusted by the committee from 16,600 bbl/day to a larger plant size of 50,000 bbl/day of oil equivalent using a scaling exponent of 0.85. This size plant is consistent with the scale used for natural gas to methanol, oil shale conversion, and direct coal liquefaction.
6. Coal to methanol to gasoline via the Mobil MTG process: Data of Schulman and Biasca (1989) modified. The size was scaled up from 16,600 bbl/day to 50,000 bbl/day, and capital investment was reduced 5 percent because of the fluid bed MTG design in comparison to the fixed bed design.
7. Shell middle distillate process: Schulman and Biasca (1989).
8. Compressed natural gas: Schulman and Biasca (1989).
9. Ethanol from corn: Schulman and Biasca (1989).
10. Oil shale pyrolysis: Schulman and Biasca (1989).
11. Tar sands pyrolysis: Schulman and Biasca (1989).
12. Solvent extraction of tar sands: Estimates made by committee based on design and capital estimates by Bechtel, Inc., using bench-scale experimental data for the process basis.
13. Direct coal liquefaction: Estimates made by committee based on design and capital estimates by Bechtel, Inc., using recent data from the Wilsonville pilot plant.
14. Other technologies: Schulman and Biasca (1989).

E

Technologies for Converting Heavy Oil

COMMERCIAL CARBON REJECTION PROCESSES

Carbon rejection processes operate at moderate to high temperatures and low pressures and suffer from a lower liquid yield of transportation fuels than hydrogen addition processes, because a large fraction of the feedstock is rejected as solid coke high in sulfur and nitrogen (and gaseous product). The liquids are generally of poor quality and must be hydrotreated before they can be used as reformer or fluid catalytic-cracking (FCC) feeds to make transportation fuels.

Delayed Coking

Heavy oil or vacuum resid is heated to above 480°C (900°F) and fed to a vessel where thermal cracking and polymerization occur. A typical product slate would be 10 percent gas, 30 percent coke, and only 60 percent liquids, the coke percentage increasing at the expense of liquid products as feeds become heavier. Since sulfur is concentrated in the coke, the coke market is limited to buyers that can control, or are not restricted by, emissions of sulfur oxides (SO_x).

Fluid Coking

Heavy oil is fed to a reactor containing a 480° to 540°C (900° to 1000°F) bed of fluidized coke particles, where it cracks to produce lighter liquids, gases, and more coke. The coke is circulated to a burner vessel where a portion of the coke is burned to supply the heat required for the endothermic coking reactions. A portion of the remaining coke is returned to the reactor as fluidizing medium, and the balance is withdrawn as product. The net coke yield is only about 65 percent of that produced by delayed coking, but the liquids are of worse quality and the flue gas from the burner requires SO_x control.

Flexicoking

Flexicoking is an extension of fluid coking. All but a small fraction of the coke is gasified to low-Btu gas (120 Btu/standard cubic foot) by addition of steam and air in a separate fluidized reactor. The heat required for both the gasification and thermal cracking is generated in this gasifier. A small amount of net coke (about 1 percent of feed) is withdrawn to purge the system of metals and ash. The liquid yield and properties are similar to those from fluid coking. The need for a coke market is eliminated or markedly reduced. The low-Btu gas can be burned in refinery furnaces and boilers or probably could also be used in cogeneration units to generate power and steam; but it must be used near the refinery since its heating value is too low to justify transportation. Unlike with fluid coking, SO_x is not an issue since sulfur is liberated in a reducing atmosphere (carbon monoxide and molecular hydrogen) inside the gasifier; however, hydrogen sulfide removal is required.

Resid FCC and Heavy Oil Cracking

This is an extension of gas oil FCC technology. Resid (usually above 650°F boiling point, not vacuum resid) is fed to a 480° to 540°C (900° to 1000°F) fluidized bed of cracking catalyst. It is converted to predominantly gasoline-range boiling materials, and the carbon residue in the feed is deposited on the catalyst. The catalyst activity is then restored by burning the deposited coke in the regenerator. This also supplies the heat required to crack the feed in the next contacting cycle. The sulfur emissions are typically controlled by additives that bind the sulfur to the catalyst for later reduction to hydrogen sulfide in the FCC reactor. The hydrogen sulfide is later processed to sulfur for sale as low-value by-product.

In resid FCC the feed contains so much carbon residue that heat exchangers (steam coils) must be installed to remove the extra heat when the added coke is burned. To be energy efficient, refineries must have an onsite use for high-pressure process steam or use it to generate electricity. Modern resid FCCs can process feeds containing up to about 10 percent carbon residue and 50 ppm of metals. This requires the use of additives to mitigate the poisoning of the catalysts by the nickel and vanadium in the feed. Since many virgin heavy oils and atmospheric resids have carbon residues of 10 to 20 percent and metals contents of 100 to 500 ppm, this process usually cannot be used as a stand-alone method for converting resid to transportation fuels. Upstream processing such as solvent deasphalting or RDS (resid desulfurization) is required.

DEVELOPMENTAL CARBON REJECTION PROCESSES WITH LIMITED COMMERCIAL DEMONSTRATION

Asphalt Residue Treatment (ART) Process

ART is a coking and vaporization process developed by the Engelhard Company, in which resid is reduced in metals and carbon residue in a prereactor similar in design to an FCC. The feed is contacted briefly with an inert solid at about 480°C (900°F) to volatilize all components of the feed except metals and carbon residue. After stripping the volatiles from the inert particles, the coke is burned off the solid in a regenerator to produce the required process heat. This is an elaborate method of distilling the feed without significant cracking. Feeds containing up to 15 percent carbon residue and 300 ppm metals can be processed, but the cost of the inert solid can become prohibitive for extremely high metals feeds. Also, flue gas desulfurization is required to handle the SO_x emissions from the regenerator. The liquid product yield is high, but quality is low and requires further upgrading. The product is suitable feed for a conventional gas oil FCC.

COMMERCIAL HYDROGEN ADDITION PROCESSES

Hydrogen addition processes include catalytic or thermal hydrocracking or donor solvent type processes. All operate at high pressure (1000 to 3000 psi) and moderate temperatures (371° to 427°C [700° to 800°F]).

Fixed Bed Residuum or Vacuum Residuum Desulfurization (RDS/VRDS)

This was originally developed over 20 years ago to remove sulfur from residual fuel oils. As available crude oils become heavier and the market for fuel oil decreases, the process is increasingly being viewed as feed pretreatment for downstream conversion units. In this process, atmospheric or vacuum residual oil contacts catalyst and hydrogen at 354° to 427°C (670° to 800°F) and about 2000 to 2500 psig, consuming about 700 to 1300 scf of hydrogen per barrel of feed. The process typically removes most of the metals and sulfur and over half of the coke precursors ("carbon residue") and hydrocracks 20 to 50 percent of the vacuum resid in the feed to primarily gas oil products. The higher conversions achievable at higher operating temperatures are not feasible with fixed bed RDS because of reactor coking and plugging. The process is not practical for extremely high metals feeds (over 250 ppm) because catalysts deactivate so fast that catalyst replacement costs are high and run lengths become prohibitively short. RDS/VRDS does not convert much heavy oil to transportation fuels directly, but it can convert many of the heaviest oils into acceptable feed for resid FCCs. Alternatively, the unconverted vacuum resid can be fed to a coker or used as a fuel oil blend stock.

Bunker Flow or Hycon Process

This process for hydrotreating resid is very similar to that for a fixed bed unit, but with the following important feature: Catalyst can be added to the top of the bed and catalyst can be withdrawn from the bottom while the unit remains onstream. This feature permits somewhat higher cracking conversions, a more uniform product slate, and longer times onstream between shutdowns.

Ebullating Bed Processes

Such catalytic hydrocracking processes known as LC-fining (developed by the Lummus Company) and H-oil (developed by Hydrocarbon Research, Inc.) can be used to demetallize, desulfurize, and hydrocrack any heavy oil. The process involves high-pressure catalytic hydrogenation but runs at higher temperatures than fixed bed RDS (about 426° to 441°C [800° to 825°F]). The feed passes upflow, expanding the catalyst bed with the ebullation and producing a back-mixed isothermal system. Reactors are very large relative to fixed bed and are frequently staged to overcome the kinetic penalties associated with back mixing.

A big advantage is the ability to add and withdraw catalyst while the unit is onstream, which allows the processing of oils with high metal concentrations than is practical with conventional fixed beds. However, catalyst replacement costs will still be high with high metals feeds. Also, the ebullating bed eliminates coke plugging problems and allows high-temperature operation and high (70 to 90 percent) vacuum resid conversions. The disadvantages include hydrogen consumption that is 20 to 100 percent higher than that for fixed bed RDS, and loss of liquid and hydrogen to high gas yields. The distillate products are low quality and require further hydrotreating and conversion to produce transportation fuels. Like fixed bed units, an economic "home" for the unreacted vacuum resid is still required.

DEVELOPMENTAL HYDROGEN ADDITION PROCESSES WITH LIMITED COMMERCIAL DEMONSTRATION

Many slurry hydrocracking processes have been developed, including CanMet, Aurabon, Veba Combicracking, and Microcat. These processes are all variations of thermal (426° to 468°C [800° to 875°F]), high-pressure (1500 to 3000 psi) hydrocracking. In a cracking reactor a dilute slurry of fine particle size, high surface area additive is present to suppress coke formation and attract feed metals, removing them from liquid products. These additives include vanadium sulfides, molybdenum sulfides, iron, and cobalt-molybdenum deposited on coke or coal. Conversions of vacuum resid are high (60 to greater than 90 percent). The uncracked bottoms are of poor quality, and at high conversions they are probably suitable only as coker feed, hydrogen plant feed, or to burn for process heat. The additives are not very active hydrogenation catalysts, so the products from the cracking reactor are fairly high in sulfur and nitrogen unless they are further hydrogenated in a second stage. Several processes have an integrated fixed bed catalyst second stage to further hydrogenate, desulfurize, and identify the cracked products with boiling points below 538°C (1000°F).

F

Retorting Technologies for Oil Shale

HOT GAS RETORTING PROCESSES

In the Paraho internal combustion retort a moving bed of shale travels downward and is heated by hot gas flowing upward through it. The hot gas is provided by combustion of hydrocarbon gas and char within the retort. Oil produced from the pyrolysis process is carried out of the retort by the gas stream in the form of vapor and oil mist (liquid). Paraho demonstrated this process at a scale of about 300 tons/day.

In the Union B and Petrosix retorts the hot gas is provided by heating the hydrocarbon gases produced by pyrolysis in an external heater. The hot gas flows countercurrent to the shale particles. In the Union B process the shale is pumped upward and the gas flows downward; in the Petrosix process the shale moves downward and the gas flows upward. Fuel from an external source is normally supplied to operate the heater. Petrosix has been operating in Brazil more or less continuously since 1972 at 2200 tons/day (800 bbl/day) in a retort 18 ft in diameter and is constructing a 7800-ton/day (3300-bbl/day) retort with a diameter of 36 ft. Union has constructed a plant to produce 10,000 bbl/day. Construction was finished in 1983, and oil shipments began in 1986. Work continues in an effort to increase the achieved capacity of 7000 bbl/day to design capacity.

In a modified in-situ process a fixed bed retort is constructed underground by some combination of mining and blasting. Downward-flowing gas is heated by burning the char in the retort. Occidental Petroleum and the U.S. Department of Energy have been the most active in developing this process, demonstrating its feasibility at Logan Wash in the early 1980s. Occidental Petroleum is proposing to demonstrate its process on its Colorado tract, C-b, where commercial-sized mine shafts have been constructed.

Rio Blanco (Gulf and Amoco) developed variations of the process in the early 1980s. Geokinetics was also active then in developing a horizontal insitu process, which avoided mining by using near-surface blasting.

HOT SOLID RETORTING PROCESSES

In the Tosco II process hot ceramic balls are mixed with smaller shale particles in a rotating drum. After the shale is pyrolyzed, the balls are separated and reheated in a ball heater using gas as fuel. The retorted shale is discarded. Tosco operated this process on a scale of 1000 bbl/day in the late 1970s and early 1980s.

In the Lurgi process hot burned shale is rapidly mixed with raw shale in a screw mixer and then held in a surge bin for a few minutes to complete pyrolysis. After pyrolysis the shale is fed to a lift pipe (dilute fluid bed), where it is burned as it is pneumatically lifted in a stream of air. The hot burned shale exiting from the lift pipe provides a continuous source of hot solid. Rio Blanco (Gulf and Amoco) operated a 1- to 5-tons/day pilot until 1984.

In the Chevron process raw shale and hot burned shale are mixed in a staged fluidized bed and held there until pyrolysis is complete. A lift pipe is used to burn the retorted shale and provide hot solid, as in the Lurgi process. Chevron constructed and briefly operated a 350-tons/day pilot plant until 1984.

In the Lawrence Livermore National Laboratory (LLNL), hot solids process mixing of recycled hot shale with raw shale occurs in a few tens of seconds in a fluidized bed. Pyrolysis is completed in a few minutes as the hot shale particles flow through a bin. The char is burned to produce hot solid in a cascading burner. Solid particles cascade down through a series of rods that slow their fall and provide sufficient residence time for combustion. Oxygen is provided by air flowing across the tumbling solid particles. The LLNL has operated a laboratory-scale retort at 1 ton/day, which is currently being enlarged to 4 tons/day.

G

Research, Development, and Demonstration Definitions

RESEARCH

Fundamental Research

This is basic research into scientific principles aimed at providing a knowledge base in chemistry and physics relevant to areas of interest in which such knowledge is not yet available.

Desired Results—Acquired knowledge may provide the basis for breakthroughs for conception and development of new technologies and programs.

Example—Understanding the chemical structure of coal might lead to less severe and more optimal reaction conditions.

Exploratory Research

Building in part on the knowledge available from fundamental research and recognizing the problem areas in process R&D, exploratory research generates new innovative technical approaches through the use of small, low-cost, bench-scale experiments. It is recognized that there is a relatively low chance of success for this type of research, but it is of low cost with occasional breakthroughs.

Desired Results—New process and catalyst concepts and insights with sufficient data on feasibility and general limitations to shape or influence process R&D.

Example—Exploratory research would try different reagents and process conditions in small laboratory tests to probe chemical structure functionalities to achieve better yields.

Catalyst Development

Building in part on fundamental catalyst characterization and kinetic mechanistic knowledge, catalyst research synthesizes and optimizes new catalyst formulations in small equipment.

Desired Results—Definition of catalyst component effects to define optimal catalyst formulations for given sets of reactants and desired products.

Example—Small laboratory batches for catalysts for process research experiments.

Process Research

Process research uses the knowledge base of fundamental and scientific disciplines to confirm practical process concepts. These concepts are usually tested in nonintegrated experiments, to show process and equipment feasibility, test operating limits, and produce data for preliminary evaluation.

Desired Results—Definition of process variable effects to define desired operating points and economics and to form the basis for an integrated pilot plant design.

Example—Batch and once-through process steps of bitumen extraction from tar sands.

DEVELOPMENT AND DEMONSTRATION

Catalyst Development

Development of commercial catalyst manufacturing techniques to provide commercial quantities of catalysts for life studies and operation in an integrated pilot plant. Pilot plant operation serves to test catalyst commercial viability.

Desired Results—Firm basis including all steps and interrelationships that can be used for economic evaluation and process design for a demonstration or commercial unit.

Example—Amocat-1A in the H-coal pilot plant.

Process Development

Development of process concepts in larger integrated pilot plants that simulate, to the extent possible, the viability of a commercial process for an extended period of operation. This is normally a more expensive phase, and a consortium may be desirable to pursue it.

Desired Result—Firm process basis including all steps and interrelationships that can be used for economic evaluation and process design for a demonstration or commercial unit.

Example—The 6-ton/day Wilsonville, Alabama, coal liquefaction pilot plant.

Process Demonstration

Demonstrate the commercial viability on equipment of "commercial size." Operability and maintenance requirements are demonstrated over very long run periods. This is a very expensive phase and should be undertaken only when technical and economic confidence in the process is high and there is sufficient industry interest to form a consortium.

Desired Results—Definition of design and economic information that can be used for (1) deciding the desirability of a commercial unit and (2) the effective construction and operation of a commercial unit when one is appropriate.

Program Area

This is defined as that portion of the program dealing with a particular technology resource combination, such as direct coal liquefaction by high-pressure hydrogenation, coal gasification, mild coal gasification, underground coal gasification, oil shale retorting, and tar sands extraction.

H

Coprocessing Technology

Signal-UOP coprocessing technology (Luebke and Humbach, 1987) has its origins in resid upgrading and differs from most of the other processes in that it is a single-stage entrained-catalyst system. In the process (Figure H-1) hydrogen, coal, petroleum resid, and catalyst are fed to a single-stage conversion reactor. After separation of the distillate products, the vacuum resid is processed to recovery catalyst for recycle. The process has been tested with bituminous and subbituminous coals. It operated well and gave resid conversions of about 55 percent. The UOP process is designed to be integrated into an existing refinery with the capability to handle the remaining resid from the coprocessing unit. Therefore, the conversion level reported above is acceptable in commercial operation. Depending on the ease of catalyst recovery, the UOP process is characterized by a simple flow sheet. The requirement of refinery integration may not be a severe limitation, because the predominant application of coprocessing is expected to be in retrofit situations.

HRI coprocessing technology provides a clear link between laboratory and pilot plant development and possible near-term commercialization. The HRI two-stage coprocessing technology grew out of the COIL process, much as HRI two-stage liquefaction technology developed from the H-coal process. Each is based on the commercial H-oil process for heavy resid upgrading. The H-oil process and similar expanded bed hydrotreaters are in commercial use in the United States, Canada, Mexico, and Kuwait. The HRI coprocessing scheme (Figure H-2) feeds a coal-oil slurry to two expanded bed catalytic reactors in series. The use of a recycle stream is optional, depending on the feed coal-to-resid ratio.

HRI has processed coals from lignite to high-volatile A bituminous (Duddy et al., 1989), achieving high conversions (greater than 90 percent at 975°F+),

APPENDIX H

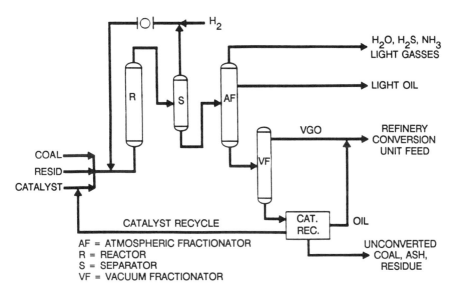

FIGURE H-1 Signal/UOP's proposed resid/coal processing scheme. SOURCE: Luebke and Humbach (1987).

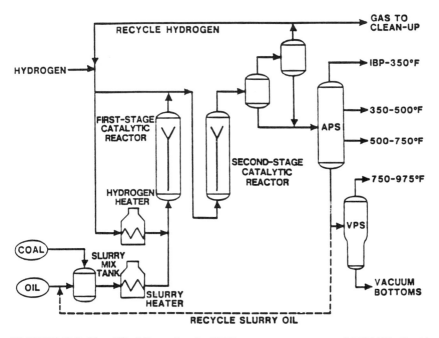

FIGURE H-2 Simplified flow plan for HRI's two-stage process. SOURCE: Duddy et al. (1986).

desulfurization (86 percent), denitrogenation (80 percent), and demetallation (99 percent). Because of the high conversions, HRI coprocessing technology could be considered for either retrofit to an existing refinery or for a stand-alone grass-roots facility. The argument that this technology could be commercialized within the 5-year study time frame is based on the commercial status of the expanded bed technology and the successful operation of the H-coal pilot plant in Cattletsburg, Kentucky, in the early 1980s. This pilot plant operated at 500 tons/day of coal and produced about 1500 bbl/day of liquids with the process configuration used. This operation demonstrated various critical components, such as slurry pumps and let-down valves, for use with coal on a precommercial scale.

I

Technical Data for Coal Pyrolysis

This appendix contains tables indicating properties of products from various coal pyrolysis processes.

TABLE I-1 Representative Operating Conditions and Yields from Recently Developed Coal Pyrolysis Processes

Process	Status	Reactor Type	Reactor Temperature (°C)	Reactor Pressure (MPa)	Residence Time(s)	Coal	Yield, Percent Dry Coal			
							Char	Tar/Oil	Gas	Water
COED, FMC Corp.	Developed	Multiple Mechanical retort	290-565	0.12-0.19	1200[a]	Bituminous	62	21	14	3
TOSCOAL	Developed	Rotating horizontal retort	520	0.1	300-600	Subbituminous	69	13	9	9
Lurghi-Ruhrgas (twin screw)	Commercial	Mechanical mixer	595	0.1	3	Subbituminous	50	32[b]	11	7[b]
Occidental flash pyrolysis	Developing	Entrained flow	610	0.3	1.5	Bituminous	56	35	7	2
Clean coke, U.S. steel	Developing	Fluidized bed	800	0.8	3000	—	66	14	15	—
Rockwell/Cities Service flash hydropyrolysis	Developing	Entrained flow	845	3.5	0.1	Bituminous	46	38	16	—

NOTE: Size of facility (metric ton/day)—COED (Char Oil Energy Development Process), 32; TOSCOAL, 23; Lurghi-Ruhrgas, 1600; Occidental flash pyrolysis, 3; Rockwell/Cities Service, 22.

[a] In pyrolysis stages.
[b] Water plus tar and oil sum 39 percent, individual values estimated.

SOURCE: Adapted from Khan and Kurata (1985).

TABLE I-2 Summary of Operating Condition Effects on Liquid Yields and Quality from Coal Pyrolysis

---

Liquid yields are maximized with high-volatile bituminous coals, and lower-rank coals yield gases and tars with higher oxygen content than bituminous coals.

The degree of primary devolatilization (gases + liquids) increases with increasing temperature, but the distribution of gases and liquids depends on secondary reactions of devolatilization products.

For a given reactor system, higher temperatures result in tars cracking to gases and secondary char, the loss of aliphatic side-chains from condensibles, a lower heteroatom content in the condensibles, a higher pitch content in the condensible product, higher BTX (benzene, toulene, xylene) yields, and lower PCX (phenols, creosols, xylenols) yields.

Char volatile matter and sulfur content decrease with increasing temperature.

Carbon oxides in the fuel gas product increase with temperature.

In inert atmosphere, condensible product yields tend to decrease with increasing pressure, but in a reactive atmosphere (water, hydrogen, carbon dioxide) condensible yields tend to increase with increasing pressure.

Reactive gases may inhibit secondary reactions of volatiles, which results in increased condensible yields.

Particle heating rates by reactor type generally follow the trend: fixed bed < moving bed < fluidized bed < entrained bed.

Slow heating favors secondary reactions of volatiles within coal particles, which results in lower overall condensible yields, more aliphatic components, a lower pitch content in the condensible product, and a composition tending to be richer in thermodynamically stable products. Fast heating favors more rapid release of tars into the gas stream, which results in higher overall condensible yields, a more highly aromatic condensible product, a higher pitch content in the tar, and a product character that tends to be determined by devolatilization kinetics.

Larger coal particles tend to favor secondary reactions within particles, producing the same results as slow heating rates.

Sulfur-capture additives such as CaO and Fe_2O_3 can reduce the H_2S content of product gases by several orders of magnitude, but they provide catalytic surfaces that tend to promote cracking reactions of condensibles. Sulfur-capture additives also reduce volatile products yield and consume carbon from the coal feed; stream pretreatment below devolatilization temperatures may enhance condensible yields.

SOURCE: Wootten et al. (1988).

TABLE I-3 Comparison of Typical Condensibles from Coalite and Occidental Flash Pyrolysis

	Process Coalite	Occidental
Property of Tar		
Run no.	N/A	175
Yield (wt% dry coal)	9.0	25.0
Specific gravity	1.029	1.143
Moisture (wt%)	2.2	(maf)
Ash (wt%)	0.1	(maf)
Elemental Analysis	*maf wt%*	
Carbon	84.0	78.8
Hydrogen	8.3	7.1
Sulfur	0.74	1.7
Nitrogen	1.08	1.8
Oxygen	5.78	10.6
Hydrogen-to-carbon atomic ratio	1.18	1.07
Toluene-insolubles	1.2	18.0
Distillation Range (°F)	*Cumulative wt%*	
<550°F	44	10
551-680°F	67	25
680°F end point	N/A	49
Total Pitch (st. percent of tar)	26.0	53.3

NOTE: maf = moisture and ash free.

SOURCE: Wootten et al. (1988).

TABLE I-4 Chars from Mild Gasification of Midwestern Coals (illustrated by chars from the third stage of FMC COED fluidized bed process)

	Illinois No. 6		W. Kentucky Nos. 9 and 14		Pittsburgh No. 8	
	Coal	Stage 3 Char	Coal	Stage 3 Char	Coal	Stage 3 Char
Moisture (wt%)	9.0[a]	0.9	3.5[a]	1.1	1.8[b]	0.2
Proximate (wt% dry)						
VM	38.5	9.3	38.0	10.9	38.3	10.6
FC	50.0	67.6	53.3	73.0	52.7	76.9
Ash	11.5	23.1	8.7	16.1	9.0	12.5
Ultimate (%)						
Carbon	67.3	63.8	70.3	74.1	73.0	75.3
Hydrogen	4.9	1.1	5.2	2.0	5.2	2.8
Nitrogen	1.3	0.8	1.5	1.6	1.3	1.4
Sulfur	4.3	6.3	3.1	3.3	4.3	4.0
Oxygen	10.7	0.0	11.2	3.0	7.2	4.0
Ash	11.5	28.0	8.7	16.0	9.0	12.5
Higher heating value (Btu/lb dry [calculated])	12,230	10,060	12,990	12,010	13,460	13,300
Density (lb/ft^3)	48.8	27.5	45.5	31.3	N/A	23.8

TABLE I-4 Continued

	Illinois No. 6		W. Kentucky Nos. 9 and 14		Pittsburgh No. 8	
	Coal	Stage 3 Char	Coal	Stage 3 Char	Coal	Stage 3 Char
Size consist Cumulative percent Tyler mesh						
14	20.2	3.7	10.1	3.5	13.2	12.2
28	57.2	40.0	35.2	35.6	42.9	50.3
48	75.9	67.9	64.0	66.7	64.6	70.5
100	86.2	83.1	73.0	85.2	79.0	81.5
200	92.2	92.4	85.1	93.5	82.6	85.6
325	95.2	96.1	92.1	97.4	93.4	93.7
Pan	99.7	99.6	100.0	100.0	100.0	100.0
Average Process Temperature (°F)	—	990	—	885	—	840

[a] As fed to Stage 1.
[b] As received.

J

Description of Technologies for Direct Conversion of Natural Gas

Numerous direct methane conversion routes are being studied at the bench scale by various companies, government agencies, and universities. These are briefly discussed in the following sections.

COLD FLAME OXIDATION

Cold flame oxidation involves the conversion of a pressurized mixture of methane and oxygen at moderate temperatures of 660° to 930°F (350° to 500°C). The reaction mixture is very fuel rich, with methane-to-oxygen ratios of about 20:1. The main chemical reaction is the oxidation of methane to methanol ($CH_4 + \frac{1}{2} O_2 = CH_3OH$). However, the further oxidation of methanol to formaldehyde often takes place simultaneously ($CH_3OH + \frac{1}{2} O_2 = CH_2O + H_2O$). The best results, achieved at the University of Manitoba, report 90 percent methanol selectivity at 7.5 percent single-pass conversion in an isothermal reactor (Kuo, 1987).

DIRECT OXIDATION

Direct oxidation involves the catalytic coupling (oxidative coupling) of methane and an oxidant in the presence of a catalyst at moderate (700°C) temperatures and at about atmospheric pressure to produce C_2+ hydrocarbons. The oxidants include air, oxygen, and nitrous oxide. Phillips Petroleum has described the use of catalysts, such as CaO/Li_2CO_3, that lead to relatively high conversions (15 to 20 percent) and good selectivities (60 to 70 percent) in the formation of ethylene and ethane. Work has been done by J. H. Lunsford and T. Ito at Texas A&M University on the chemistry of direct oxidation. Also, Akzo Chemie, Amoco, University of California at

Berkeley, the universities of Pittsburg and Tokyo, Idemitsu Kosan, and Union Carbide have also pursued development of direct-oxidation catalysts.

British Petroleum has described experiments at higher temperatures (1100°C) and at short residence times at which conditions methane reacts with oxygen to produce syngas as well as C_2+ hydrocarbons. Several catalysts, including zirconia, gave C_2+ selectivities of over 50 percent.

OXYCHLORINATION

Oxychlorination involves the catalytic reaction of methane with a mixture of hydrogen chloride and oxygen to produce methyl chloride. The methyl chloride is then reacted over a shape-selective zeolite catalyst to produce a mixture of aliphatic and aromatic hydrocarbons. Methane can also be oxidized to methyl halides using chlorine, bromine, or iodine.

British Petroleum has developed catalysts that are selective in the conversion of methane and also the conversion of methyl chloride to hydrocarbons. The Pittsburgh Energy Technology Center has announced a similar process. The chloromethanes are reacted over ZSM-5 to produce gasoline and HCl, which is recycled. Imperial Chemical and Mobil have described catalysts that support the oxychorination process.

INDIRECT OXIDATION (OXIDATIVE COUPLING TO ETHYLENE)

Indirect oxidation of methane takes place at high temperature (about 760°C) using various reducible metal oxides as oxygen carriers as well as catalysts. A typical reaction can be represented by

$$2\ CH_4 + MOx + 2 = C_2H_4 + 2\ H_2O + MOx,$$

where $MOx + 2$ and MOx represent the metal oxide and reduced metal oxide, respectively. The reducible oxides are reoxidized with oxygen according to the reaction $MOx + O_2 = MOx + 2$. Combination of the above reactions gives the following "coupling" reactions:

$$CH_4 + \tfrac{1}{2}\ O_2 = \tfrac{1}{2}\ C_2H_4 + H_2O.$$

The indirect oxidation route can use air rather than purified oxygen and operates at a temperature that allows efficient recovery of the reaction heat. This is done by circulation of the metal oxide between a zone where it reacts with methane and a reoxidation chamber where it is regenerated with air. This configuration is similar to a fluid bed catalytic cracking unit, in which the catalyst circulates between a cracking section and a regeneration section (where coke is burned off the cracking catalyst). To make a liquid fuel, the oxidative coupling product would be cascaded over a reactor used in the methanol to gasoline process to oligomerize the ethylene to gasoline.

ARCO Oil and Gas Company currently appears to be the leader in indirect oxidation, with 45 patents. ARCO recently reported success in developing a two-step process that first converts methane to olefins (called REDOX process), followed by a catalytic reaction of the olefins to high-octane gasoline. In the first step, ARCO has reported conversions of 25 percent with C_2+ selectivities greater than 75 percent. According to ARCO, this technology has been proven in pilot-scale studies and is ready for larger-scale testing in a demonstration plant.

CATALYTIC PYROLYSIS

Direct methane conversion through catalytic pyrolysis involves contact of methane with a catalyst at a relatively high temperature (1100° to 1200°C), pressures near about 1 atm, and at a short contact time. Under these conditions methane undergoes catalytic dehydrogenation ($2\ CH_4 = C_2H_4 + 2\ H_2$). Chevron has described several catalysts, such as $Al_2O_3/Th/Cs$, that lead to relatively high conversions (20 percent) with high selectivity (90 to 100 percent) to C_2+ hydrocarbons, including both light olefins and aromatic hydrocarbons. The University of Houston, Phillips Petroleum, Sohio, and the University of Taiwan have also pursued catalytic pyrolysis.

STRONG ACID CONVERSION

Strong acid catalysts can promote polycondensation of methane. The University of California, Exxon Corporation, and Firestone have investigated this chemistry.

BIOLOGICAL CONVERSION

Biological conversion utilizes organisms that consume methane and oxygen for growth, thereby producing methanol. Exxon, several Japanese institutions, and the University of Michigan have reported work on the biological conversion route.

K

Temperature Characteristics of High-Temperature Gas Reactors

The high-temperature gas-cooled reactor (HTGR) is promising for the production of high-temperature process heat. The present version has an outlet coolant temperature of about 700°C and can be applied to the generation of electricity and steam at high pressure. The high pressure facilitates the transport of steam to remote locations, which potentially increases the application of nuclear energy to enhanced oil recovery operations, and to the "mining" of heavy oils and tar sands.

In the future the HTGR outlet coolant temperature can be increased (eventually to 950°C and higher), so that HTGRs become very high temperature reactors (VHTRs) and become a potential source of high-temperature process heat. VHTRs show potential for application to fossil fuel conversion processes (e.g., steam reforming of methane at 850°C beyond the year 2005 and steam gasification of coal at higher temperatures beyond the year 2010). At 900°C and above, steam gasification of coal produces synthesis gas for the production of transportation fuels. The use of VHTRs as the energy source in coal gasification could reduce coal use by about 35 percent relative to using coal as the process energy source.

VHTR use in the reforming of methane to synthesis gas could reduce natural gas use by about 40 percent relative to using natural gas as the process energy source.

Glossary

AAPG. American Association of Petroleum Geologists.
AOSTRA. Alberta Oil Sands and Tar Research Authority.
API. American Petroleum Institute.
API gravity. Parameter expressed in degrees where the specific gravity of water is defined as 10°; light oil is 20° and greater; heavy oil ranges from 10° to 20°, and tar sands range from 0° to 10°.
Aromatic hydrocarbons. Compounds or mixtures of compounds having benzene rings in their molecules; useful as antiknock additives for gasolines.
ART. Asphalt residue treatment.

bbl. Barrel(s).
Bcf. Billion cubic feet.
Bitumen. Also known as asphalt; found in tar sands deposits in Utah and elsewhere.
Btu. British thermal unit.
Bunker flow process. Process for hydrotreating residium.

CANMET. Canada Centre for Mineral and Energy Technology.
CARB. California Air Resources Board.
C_i. Number of carbon atoms in a hydrocarbon molecule where i is the number; for example, C_1-C_3 refers to molecules with one to three carbon atoms, the lighter gaseous components of a product.
Catalytic cracking. Cracking process in which a catalyst is used to facilitate the reaction.
CFCs. Chlorofluorocarbons.
COED. FMC's char oil energy development process.
CNG. Compressed natural gas.

201

Coke. Hard solid residue left after coal has undergone carbonization and has its volatile matter (oils and gases) removed.

Cracking. Process by which hydrocarbons are decomposed by thermal or catalytic means to produce lower-boiling fractions suitable for chemical feedstock or gasoline.

DCF. Discounted cash flow.

Distillate fuel. General term used to describe fuels obtained by the fractional distillation of crude petroleum.

DOE. U.S. Department of Energy.

Ebullating bed reactor. In this design the upward linear velocity of hydrogen and hydrocarbons is sufficiently high to expand the bed of catalyst particles and induce a continuous random motion.

EDS. Exxon donor solvent process for the liquefaction of coal in which a solvent is hydrogenated, mixed with coal to form a slurry, and then fed to a liquefaction reactor.

EIA. Energy Information Administration.

Endothermic. Characterized by or formed with the absorption of heat.

EOR. Enhanced oil recovery; method to increase ultimate oil production beyond that achieved by primary and secondary methods.

Exothermic. Characterized or formed with evolution of heat.

FCC. Fluid catalytic cracking.

Fischer assay. Standardized test that measures the amount of liquid oil that can be obtained from ordinary pyrolysis processes.

F-T. Fischer-Tropsch.

Fischer-Tropsch process. Catalytic conversion of synthesis gas into a range of hydrocarbons.

Fluidized Bed processes. Processes that rely on the tendency of finely divided solids to float in a low-velocity gas or liquid stream and behave as a fluid.

Heteroatom. An atom other than carbon in the ring of a heterocyclic compound.

H-Oil. Ebullating bed, catalytic hydrocracking process developed by Hydrocarbon Research, Inc.

HRI. Hydrocarbon Research, Inc.

HTGR. High-temperature gas-cooled reactor.

Hycon process. Process for hydrotreating residuum.

Hydrocracking. Catalytic cracking of higher petroleum fractions in the presence of hydrogen to produce fractions with a lower boiling point.

IGCC. Integrated gasification combined cycle.

Isomerization. Process in which a straight-chain hydrocarbon is converted into its branched-chain analog.

LC-fining. Ebullating bed, catalytic hydrocracking process developed by Lummus Company.
LLNL. Lawrence Livermore National Laboratory.
Mcf. Thousand cubic feet.
MDS. Middle distillate synthesis.
METC. Morgantown Energy Technology Center.
MJ. Megajoule.
MIS. Modified in situ.
MMbbl. Million barrels.
MMBtu. Million British thermal units.
MMS. Minerals Management Survey.
MOGD. Mobil olefins to gasoline and diesel process.
mpg. Miles per gallon.
MTG. Mobil methanol to gasoline process.
MTO. Methanol to olefins process.

Naptha. Sometimes known as "heavy gasoline" or light distillate feedstock, it is the petroleum fraction boiling in the 70° to 200°C range.
NG. Natural gas.
NO_x. Oxides of nitrogen.
NPC. National Petroleum Council.

Oil shale. Carbonaceous rock containing a high-molecular-weight polymer called kerogen that can produce oil when heated to pyrolysis temperatures.
Olefin. Alternative name for alkenes, aliphatic hydrocarbons whose molecules contain a double bond and have the general formula C_nH_{2n}; olefins are more reactive than paraffins (alkanes), have high octane numbers, and are therefore blended into motor gasolines.
O&M. Operation and maintenance.
OOIP. Original oil in place.
OPEC. Organization of Petroleum Exporting Countries.

Partial oxidation. Net effect of a number of component reactions that occur in a flame supplied with less than stoichiometric oxygen.
PERF. Petroleum Environmental Research Forum.
PGC. Potential Gas Committee.
Pyrolysis. Thermal decomposition of a chemical compound or mixture of compounds.

RAPAD. Research Association of Petroleum Alternative Development.
R&D. Research and Development.
RD&D. Research, development, and demonstration.
RDS. Residuum (resid) desulfurization.
RDS/VRDS. Residuum or vacuum residuum desulfurization.

REDOX process. ARCO process for direct methane conversion.

Reforming. General term for a number of secondary refining processes. It is used to increase the proportion of particular constituents in a distillate, to introduce a compound that is absent, or to improve the ignition quality.

Reserves. Amount of a resource believed to be economically recoverable with existing technology.

RHCs. Reactive hydrocarbons.

Sasol. South African Coal, Oil, and Gas Corporation; coal conversion plant in operation at Sasolburg; coal is gasified by the Lurgi process and then converted to liquid hydrocarbons through the Fisher-Tropsch process.

scf. Standard cubic foot.

SMDS. Shell middle distillate synthesis.

SO_x. Oxides of sulfur.

Synthesis gas. Mixture of carbon monoxide and hydrogen and other liquid and gaseous products.

SCT. Short contact time.

SRC. Solvent-refined coal.

Tar. Heavy, dark liquid residues obtained from petroleum refining, also called bitumen; the hydrocarbon deposits in tar sands.

Tcf. Trillion cubic feet.

Thermal cracking. Subjection of heavy distillate and residues to high temperatures and pressures to produce gasoline and gas oil; visbreaking and delayed coking are thermal cracking processes.

TIGAS. Topsoe Integrated Gasoline Synthesis.

UCG. Underground coal gasification.

USGS. U.S. Geological Survey.

Vacuum residue. Residue left behind after vacuum distillation; consists of the heaviest fuel oils and bitumens.

VHTR. Very high temperature reactor.

VOC. Volatile organic carbon.

WRI. Western Research Institute.

References and Bibliography

Alson, J., J. Adler, and T. Baines. 1989. The Motor Vehicle Emission Characteristics and Air Quality Impacts of Methanol and Compressed Natural Gas. In Alternative Transportation Fuels: An Environmental and Energy Solution, D. Sperling, ed. Westport, Conn.: Quorum Book/Greenwood Press.

American Petroleum Institute (API). 1989. Monthly Statistical Report, Vol. 13, No. 5 (May). Washington, D.C.

American Petroleum Institute (API). 1986. API Basic Petroleum Data Book: Petroleum Industry Statistics, Vol. 6, Section 2, Table 3c; Section 4, Table 2c. Washington, D.C.

American Association of Petroleum Geologists (AAPG). 1989a. Position Paper on U.S. Oil Resources. Explorer, Vol. 10, No. 9, Tulsa, Okla.

American Association of Petroleum Geologists (AAPG). 1989b. Position Paper on U.S. Natural Gas Resources. Explorer, Vol. 10, No. 9, Tulsa, Okla.

Ashland Synthetic Fuels, Inc., and Airco Energy Company, Inc. 1981. The Breckenridge Project—Initial Effort, Reports I through X. DOE/OR/20717. Washington, D.C.: U.S. Government Printing Office.

Bechtel, Inc. 1989. Western States Enhanced Oil Shale Recovery Program—Shale Oil Production Facilities. Conceptual Design Studies Report (August). Bechtel Job Number 20227.

Bergius, F. 1913. German Patent 301.231.

Beyaert, B. K., S. K. Hoekman, A. J. Jessel, J. S. Welstand, R. D. White, and J. E. Woycheese. 1989. An Overview of Methanol Fuel Environmental, Health and Safety Issues. Symposium on Alternative Transportation Fuels for the 1990s and Beyond. Philadelphia, Pa.: American Institute of Chemical Engineers.

Brown, A., and B. M. Stewart. 1978. Water Management for Oil Shale Mining and Retorting in the Piceance Creek Basin, Colorado. Eleventh Oil Shale Symposium Proceedings, sponsored by Colorado School of Mines and Laramie Energy Research Center, Golden, Colo.

Bureau of Economic Geology (BEG). 1988. An Assessment of the Natural Gas Resource Base of the United States. Report of Investigations No. 179, University of Texas, Austin.

California Fuel Methanol Study. 1989. Available from Chevron USA, Inc., Strategic Planning and Business Evaluation, San Francisco, California.

Carter, W. P. L., et al. 1986. Effects of Methanol Substitution on Multi-day Air Pollution Episodes. Statewide Air Pollution Research Center, University of California at Riverside. Under Contract to California Air Resources Board, ARB-86: meth 86.

Cena, R. J., and R. G. Mallon. 1986. Results and Interpretation of Rapid-Pyrolysis Experiments Using the LLNL Solid-Recycle Oil Shale Retort. Nineteenth Colorado Oil Shale Proceedings, Colorado School of Mines (August).

Cena, R. J., C. B. Thorsness, and J. A. Britten. 1988. Assessment of the CRIP Process for Underground Coal Gasification: The Rocky Mountain I Test. Presented at the AIChE Summer National Meeting, Denver, Colo., August 22-24. Lawrence Livermore National Laboratory No. UCRL 98929.

Coal and Synfuels Technology. 1989a. Arlington, Va.: Pasha Publications, p. 5 (February 6).

Coal and Synfuels Technology. 1989b. Arlington, Va.: Pasha Publications, p. 3.

Considine, D. M. (ed.) 1977. Energy Technology Handbook. New York: McGraw-Hill.

DeSlate, E. 1984. Technical and Economic Assessment of the Occidental Coal Flash Pyrolysis Process. Document AP 3786 (December). Palo Alto, Calif: Electric Power Research Institute.

DeLuchi, M. 1989. Greenhouse Gases from Alternate Fuel Vehicles. Final Report, June 10. Argonne, Ill.: Center for Transportation Research, Argonne National Laboratory.

DeLuchi, M. 1988. Hydrogen Vehicle: An Evaluation of Fuel Storage, Performance, Safety, Environmental Impacts, and Cost. International Journal of Hydrogen Energy 14(2):81-130.

DeLuchi, M., R. A. Johnston, and D. Sperling. 1988a. Transportation Fuels and the Greenhouse Effect. Transportation Research Record 1175: 33-44.

DeLuchi, M., R. A. Johnston, and D. Sperling. 1988b. Methanol vs. Natural Gas Vehicles: A Comparison of Resource Supply, Performance, Emissions, Fuel Storage, Safety, Costs, and Transitions. SAE Techni-

cal Paper Series 881656. Warrendale, Pa.: Society of Automotive Engineers.
Dickie, B., and M. Carrigy. 1977. Fuel from Tar Sands, In Energy Technology Handbook, D. M. Considine, ed. New York: McGraw-Hill.
Duddy, J. E., J. B. McArthur, and J. B. McLean. 1986. HRI's Coal/Oil Coprocessing Program—Phase I. Proceedings, 11th Annual EPRI Contractor's Conference on Clean Liquid and Solid Fuels. Palo Alto, Calif.: Electric Power Research Institute.
Duddy, J. E., J. B. McLean, and T. O. Smith. 1989. Coal/Oil Coprocessing Program Update. Proceedings, 12th Annual EPRI Contractor's Conference on Fuel Science. Palo Alto, Calif.: Electric Power Research Institute.
Dunker, A. M. 1989. The Relative Reactivity of Emissions from Methanol-Fueled and Gasoline-Fueled Vehicles in Forming Ozone. Presented to the Air and Waste Management Association, Anaheim, Calif. June 25-30.
Ember, L. R., P. L. Layman, W. Lepkowski, and P. S. Zurer. 1986. Tending the Global Commons. Chemical and Engineering News (November 24):14-64.
Energy Information Administration (EIA). 1984. Annual Energy Review. DOE/EIA-0384(84). Washington, D.C.: EIA.
Energy Information Administration (EIA). 1987a. Performance Profiles of Major Energy Producers. DOE/EIA-0206(87). Washington, D.C.: U.S. Government Printing Office.
Energy Information Administration (EIA). 1987b. International Energy Annual 1987. DOE/EIA-0219. Washington, D.C.: U.S. Government Printing Office.
Energy Information Administration (EIA). 1988. International Energy Outlook 1987. DOE/EIA-0219, Table 25, p. 78 (October). Washington, D.C.: U.S. Department of Energy.
Energy Information Administration (EIA). 1989a. Annual Energy Outlook with Projections to 2000. Washington, D.C.: U.S. Department of Energy.
Energy Information Administration (EIA). 1989b. Annual Outlook for U.S. Coal. 1989. DOE/EIA-0333(89). Washington, D.C.: U.S. Government Printing Office.
Energy Information Administration (EIA). 1989c. International Energy Outlook 1989: Projection to 2000. Washington, D.C.: U.S. Government Printing office.
Energy Information Administration (EIA). 1989d. Monthly Energy Review (January). Washington, D.C.: U.S. Government Printing Office.
Fant, B. T. 1973. How New Technology Can Reduce Process Costs. Paper presented at the 74th National AIChE Meeting, New Orleans, March 11-15.

Fluor Corporation. 1988. An Analysis of the Cold Flame Oxidation Route. Paper presented at Session 70b, Fuels and Chemicals from Natural Gas. AIChE Convention (Spring), New Orleans, La.

Gessner, A. W., T. J. Hand, and J. M. Klara. 1988. Mild Gasification of Steam Conditioned Bituminous Coal in Fluidized Beds. Technical Note DOE/METC-88/4085. Washington, D.C.: U.S. Department of Energy.

Greene, M., A. Gupta, and W. Moon. 1986. American Chemical Society Division of Fuel Chem Prepr. 31(4):208-215.

Harris, J. N., A. G. Russell, and J. B. Milford. 1988. Air Quality Implications of Methanol Fuel Utilization. Future Transportation Technology Conference and Exposition. SAE Technical Papers Series 881198. San Francisco, Calif.

Hu, W. C. S., and R. C. Rex, Jr. 1988. Economics of HYTORT Processing for Six Eastern Oil Shales, Eastern Oil Shale Symposium, November 30-December 2, University of Kentucky, Institute of Mining and Minerals Research, Lexington.

Hubbert, M. K. 1962. Energy Resources: A Report to the Committee on Natural Resources. Publication 1000-D. Washington, D.C.: National Academy of Sciences.

Huber, D. A., Q. Lee, R. L. Thomas, K. Frye, and G. Rundins. 1986. American Chemical Society Division of Fuel Chem. Prepr. 31(4): 227-233.

Interstate Oil Compact Commission (IOCC). 1982. Tar Sands, D. Ball, L. C. Marchant and A. Goldberg, eds., IOCC Monograph Series. Oklahoma City, Okla.: IOCC.

Interstate Oil Compact Commission (IOCC). 1984. Major Tar Sand and Heavy Oil Deposits of the United States. Oklahoma City, Okla.: IOCC.

Japan Chemical Week, Vol. 27, Issue No. 1359, p. 1, April 17, 1986.

Johanson, E. S., and A. G. Comolli. 1978. PDU Run 5 Syncrude Mode Operation with Catalyst Addition and Withdrawal. DOE Report DF/FE 2547-19. Washington, D.C.

Journal of Petroleum Technology, October, 1988. p. 1294.

Kastens, M. L., L. I. Hirst, and C. C. Chaffee. 1949. Liquid fuels from coal. Ind. Eng. Chem. 41:870-885.

Kelly, J. F., and S. A. Fouda. 1984. CANMET Coprocessing: An Extension of Coal Liquefaction and Heavy Oil Hydrocracking Technology. Proceedings, DOE Direct Liquefaction Contractors' Review Meeting. Albuquerque, N.M. Washington, D.C.: U.S. Department of Energy.

Khan, M. R., and T. M. Kurata. 1985. The Feasibility of Mild Gasification of Coal: Research Needs. Technical Note DOE/METC-85/4019. Washington, D.C.: U.S. Department of Energy.

Klara, J. M., and T. J. Hand. 1989. Mild Gasification: A New Option. Presented at Alternative Energy '89, Tucson, Ariz. April 16-18.

Kuo, J. C. W. 1984. Gasification and Indirect Liquefaction. Chap. 5 in Science and Technology of Coal and Coal Utilization, B. R. Cooper and W. A. Ellingson, eds. New York: Plenum Press.

Kuo, J. C. W. 1987. Evaluation of Direct Methane Conversion to Higher Hydrocarbons and Oxygenates. DOE-sponsored report contract no. DE-AC22-86PC90009. Mobil Research and Development Corporation.

Kuuskraa, V. A., 1985. Major Tar Sands and Heavy Oil Deposits of the United States. Proceedings of the Third UNITAR/UNDP International Conference on Heavy Crude and Tar Sands, Long Beach, Calif. July 22-31. Edmonton, Canada: Alberta Oil Sands, Technology and Development Authority.

Kuuskraa, V., K. S. McFall, and M. L. Godec. 1989. U.S. Petroleum Resources and Natural Gas Reserves. Report prepared for the Committee on Production Technologies for Liquid Transportation Fuels. Fairfax, Va.: ICF Resources, Inc.

Leubke, C. P., and M. J. Humbach. 1987. Continuous Bench-Scale Testing of Coprocessing. Proceedings, 12th Annual EPRI Contractor's Conference on Fuel Science. Palo Alto, Calif.: Electric Power Research Institute.

Lewis, A. E. 1980. Oil from Shale: The Potential, The Problems, and a Plan for Development. Energy 5:373-387.

Long, W. D., L. L. N. Parker, and M. Dodd. 1986. Effects of Methanol Fuel Substitution on Multi-day Air Pollution Episodes. Riverside, Calif.: Statewide Air Pollution Research Center.

Lumpkin, R. E. 1988. Recent Progress in the Direct Liquefaction of Coal. Science 239:873-877.

McLean, J. B., and J. E. Duddy. 1986. American Chemical Society Division of Fuel Chemical Prepr. 31(4): 169-180.

Megill, R. E. 1989. Figures Recall the Industry's Crash. AAPG Explorer (June):32.

Moses, D., and C. Saricks. 1987. A Review of Methanol Vehicles and Air Quality Impacts. SAE Fuels and Lubricants Meeting and Exposition, Toronto, Ontario, Canada, November 2-5.

National Alcohol Fuel Commission. 1981. Fuel Alcohol. Final Report, Washington, D.C.: U.S. Government Printing Office.

National Coal Association. 1986. Facts About Coal. Washington, D.C.: NCA.

National Petroleum Council (NPC). 1988. Integrating R&D Efforts. Washington, D.C.: NPC.

National Research Council (NRC). 1977. Assessment of Technology for the Liquefaction of Coal. Ad Hoc Panel on Liquefaction of Coal, Committee on Processing and Utilization of Fossil Fuels, Commission on Sociotechnical Systems. Washington, D.C.: National Academy of Sciences.

National Research Council (NRC). 1979. Hydrogen as a Fuel. Energy Engineering Board. Washington, D.C.: National Academy of Sciences.

Nichols, R. J., E. L. Clinton, E. T. King, C. S. Smith, and R. J. Wineland. 1988. A View of Flexible Fuel Vehicle Aldehyde Emissions. SAE Paper 881200: Warrendale, Pa.:

Nichols, R. J., and J. M. Norbeck. 1985. Assessment of Emissions from Methanol-fueled Vehicles: Implications for Ozone Air Quality. Paper presented at the 78th Annual Meeting of the Air Pollution Control Association, Detroit, Mich.

Oak Ridge National Laboratory (ORNL). 1987. Transportation Energy Data Book, Oak Ridge, Tenn.: ORNL.

Ogden, J., and R. Williams. 1989. Hydrogen and the Revolution in Amorphous Silicon Solar Cell Technology. Pu/CEES 231 (February 15). Princeton, N.J.: Princeton University, Center for Energy and Environmental Studies.

Oil and Gas Journal. 1986. Fiscal 1985 Returns for OGJ 400 Mixed. (September 8):55-90.

Piper, E. M., and O. C. Ivo. 1986. The Petrosix Project in Brazil—An Update. 19th Oil Shale Symposium Proceedings, August. Golden: Colorado School of Mines Press.

Rahimi, P. M., S. A. Fouda, and J. F. Kelly. 1987. Effect of Coal Concentration on Product Distribution in CANMET Coprocessing. Proceedings, 12th Annual EPRI Contractors' Conference on Fuel Science. Palo Alto, Calif.: Electric Power Research Institute.

Reed, T. B., and M. S. Graboski. 1989. Methanol Production from Biomass and Municipal Waste: Process Design and Economics. Presented to Committee on Liquid Transportation Fuels, National Research Council, Washington, D.C. (June 9, 1989). Available from T. B. Reed, Colorado School of Mines, Golden, Colorado.

Rhodes, D. E. 1985. Comparison of Coal and Bitumen-Coal Process Configurations. Proceedings, 10th Annual EPRI Contractors' Conference on Clean Liquid and Solid Fuels. Palo Alto, Calif.: Electric Power Research Institute.

Riva, J. P., Jr. 1987. Fossil Fuels. In Encyclopedia Britannica, Vol. 19, pp. 588-612.

Riva, J. P., Jr. 1988. Oil Distribution and Production Potential. Oil and Gas Journal 86(3):58.

Riva, J. P., Jr. 1989. Domestic National Gas Production. CRS Issue Brief, Congressional Research Service. The Library of Congress (May 2).

Russell, M. 1988. Tropospheric Ozone and Vehicular Emissions. ORNL/TM-10908. Oak Ridge, Tenn.: ORNL.

Schearer, S. H. 1973. Char Oil Energy Development Process. Chemical Engineering Progress. 69:(3).

Schindler, H. D., 1989. Coal Liquefaction: A Research & Development Needs Assessment. Prepared for the U.S. Department of Energy, Office of Energy Analysis, Office of Program Analysis. Report DOE/ER-0400. Washington, D.C.: U.S. Department of Energy.

Schulman, B., and F. Biasca. 1989. Liquid Transportation Fuels from Natural Gas, Heavy Oil, Coal, Oil Shale and Tar Sands: Economics and Technology. Mountain View, Calif.: SFA Pacific, Inc.

Schulman, B. L., et al. 1988. Assessment of Coprocessing of Coal and Residual Oil in the United States Refining Industry. Prepared for the U.S. Department of Energy, Assistant Secretary of Fossil Energy, Report DOE/FE/60457-H3. Washington, D.C.: U.S. Department of Energy.

Schumacher, W. J. 1989. The Economics of Alternative Fuels. SRI International Project No. 6998. Menlo Park, Calif.: SRI International.

Shinn, J. H., et al. 1984. The Chevron Co-refining Process. Proceedings, 9th Annual EPRI Contractors' Conference on Coal Liquefaction. Palo Alto, Calif.: Electric Power Research Institute.

Sierra Research, Inc. 1989. Potential Emissions and Air Quality Effects of Alternative Fuels. SR 88-11-02. Sacramento, Calif.: SRI.

Singleton, A. H. 1982. An Industrial Viewpoint of Synthetic Fuels. Proceedings of the Eighth Underground Coal Conversion Symposium, August, Keystone, Colo., sponsored by the U.S. Department of Energy.

Sparks, F. L. 1974. Water Prospects for the Emerging Oil Shale Industry. Proceedings of the Seventh Oil Shale Symposium, Colorado School of Mines, Golden, Colo., April 18-19.

Sperling, D. 1988. New Transportation Fuels: A Strategic Approach to Technological Change. Berkeley: University of California Press.

Sperling, D., and M. A. DeLuchi. 1989. Is Methanol the Transportation Fuel of the Future? Energy 14(8):469-482.

Tesche, T. W. 1988. Accuracy of Ozone Air Quality Models. Journal of Environmental Engineering, 114(4).

Tsaros, C. L., V. L. Arora, and K. Burnham. 1975. The Manufacture of Hydrogen from Coal. SAE Technical Paper 751095. Warrendale, Pa.: Society of Automotive Engineers.

Unnasch, S., C. B. Moyer, and M. D. Jackson. 1986. Emission Control Options for Heavy-duty Engines. SAE Paper 861111, Warrendale, Pa.: Society of Automotive Engineers.

U.S. Congress. 1987. U.S. Oil Production: The Effect of Low Oil Price. Office of Technology Assessment, OTA-E-348. Washington, D.C.: U.S. Government Printing Office.

U.S. Congress. 1988a. Conference Report Accompanying Public Law 100-446. U.S. Department of the Interior and Related Agencies Appropriations Act of Fiscal Year 1989. Washington, D.C.

U.S. Congress. 1988b. Alternative Motor Fuels Act of 1988. 100th Congress, 2nd Session. U.S. House of Representatives, Washington, D.C., Report 100-929.

U.S. Department of Energy (U.S. DOE). 1986. Feasibility of Establishing and Operating a Generic Oil Shale Test Facility. DOE/METC-86/6046, Morgantown, W. Va.: U.S. DOE.

U.S. Department of Energy. 1988. Assessment of Costs and Benefits of Flexible and Alternative Fuel Use in the Transportation Sector. Washington, D.C.: U.S. DOE.

U.S. Department of Energy. 1989a. Fiscal Year 1990 Congressional Budget Request, Fossil Energy Research and Development, Vol. 4. Washington, D.C.: U.S. DOE.

U.S. Department of Energy. 1989b. Energy Technologies and the Environment; Environmental Information Handbook. Assistant Secretary for Environment, Safety, and Health. Washington, D.C.: U.S. DOE.

U.S. Department of Energy, Western States Enhanced Oil Shale Recovery Program. 1989c. Modified-In-Situ (MIS) Oil Shale Project with Hot-Recycled Solid (HRS) Shale Retort. Washington, D.C.: U.S. DOE.

U.S. Department of the Interior (USDOI). 1989. Estimates of Undiscovered Conventional Oil and Gas Resources in the United States—A Part of the Nation's Endowment. Denver, Colo.: U.S. Geological Survey.

Wang, Q., M. A. DeLuchi, and D. Sperling. 1989. Air Pollutant Emissions and Electric Vehicles. Research Report UCD-TRG-RR-89-1. Transportation Research Group, University of California, Davis.

Whitehurst, D. D. 1978. A Primer on Chemistry and Constitution of Coal. Pp. 1-35 in Organic Chemistry of Coal, American Chemical Society Symposium Series No. 71. Washington, D.C.: ACS.

Whitten, G. Z., N. Yonkow, and T. C. Myers. 1986. Photochemical Modelling of Methanol—Use Scenarios in Philadelphia. EPA 460/3-86-001. Ann Arbor, Mich.: U.S. Environmental Protection Agency, Office of Mobile Sources.

Wootten, J. M., R. G. Duthie, R. A. Knight, M. Onischak, S. P. Babu, and W. G. Bair. 1988. Development of an Advanced, Continuous Mild Gasification Process for the Production of Co-products. Topical Report for the Period September 30, 1987-January 31, 1988, to U.S. Department of Energy. Work performed under Contract No. DE-AC21-87MC24266.

Yurko, W. J., and R. W. Luhning. 1988. AOSTRA Technology Development for Alberta Oil Sands and Heavy Oil. Proceedings of the Fourth UNITAR/UNDP Conference on Heavy Crude and Tar Sands, No. 243. Edmonton, Canada: Alberta Oil Sands, Technology and Research Authority.

Index

A

Academic research, 69, 98–99, 197–198, 199
Accidents and accident prevention, 31–32
Additives, asphalt, 13
Advanced Coal Liquefaction R&D Facility, 93
Advanced Mitsubishi Synthesis Gas to Gasoline, 88
Air pollution, see Environment and pollution
Air Products and Chemicals, 88
Air transport, 13
Akzo Chemie, 197
Alabama, 16, 141–143
 Wilsonville coal liquefaction project, 93, 94, 95, 96, 128, 187
Alaska, 16, 77, 91, 141
Alberta Oil Sands and Tar Research Authority, 73
Alcohol fuel, see Ethanol; Methanol
Alternative fuel sources, 2, 3, 39, 123
 cost factors, 40–56, 146–177
 environmental impacts, 1, 5, 6, 7, 14, 19, 105–114, 116, 122–123, 127
 see also specific fuels

American Association of Petroleum Geologists, 25–27
Ammonia, 63, 68
Amoco, 13, 76, 90, 93, 184, 197
ARCO Oil and Gas Company, 102, 103, 199
Aromatic hydrocarbons, 94
Ash, 62, 95, 101
Asphalt, see Bitumen
Asphalt residue treatment, 67, 180
Australia, 51–52, 93–94
Automobiles
 alternative-fuel vehicle demonstrations, 105
 catalytic converters, 107–108
 compressed natural gas as fuel, 42, 45, 53, 54–56, 107–109, 110, 111–113, 119, 150
 electric, 13
 emissions, 105–114, 122–123, 127
 methanol as fuel, 49, 50–51, 53–54, 55–56, 149, 150
 multifuel, 53–54, 107, 112, 149
 operating costs, 49, 50–51, 53–55

B

Benzene, 94
Biological conversion, natural gas, 199

Biomass energy, 4, 8, 114, 123, 126, 128–129
  DOE programs, 17, 65, 66, 126, 128–129
  environmental factors, 113–114, 128
  gasification, 65, 66
  resource base, 17, 128, 138, 145
  wood, 44–45, 111
Bitumen, 71, 73, 74, 75, 76, 120
  asphalt additive, 13
  asphalt residue treatment, 67, 180
  resources, 16, 70, 141–143
  see also Tar and tar sands
Brazil, 143
British Petroleum, 90

C

California, 16, 61, 91, 104, 105, 106, 110, 141, 143, 197–198, 199
California Air Resources Board, 105
Canada, 70, 71, 73, 76, 128, 143, 197
Canada Centre for Mineral and Energy Technology, 97–98
Cancer, 94
Capital investments, 2, 45, 48, 49, 51, 52, 68, 88, 95, 118, 123, 146–147, 150, 151–154
Carbon dioxide, 4, 8, 15, 111–114, 118, 122–123, 127, 128
  coal-oil coprocessing, 99
  heavy oil, 69, 128, 178–181
  oil shale, 78, 82–83, 127
  synthesis gas, 60, 62, 65, 66
Carbon monoxide, 106, 119
Catalysis and catalytic cracking, 66–67, 68
  auto catalytic converters, 107–108
  coal gasification, 62
  coal liquefaction, 93–94, 127
  coal-oil coprocessing, 98, 100
  Fischer-Tropsch process, 57, 89–90, 91, 92, 120, 123, 129
  fluid catalytic cracking, 66–67, 68, 179–180
  heavy oil, 68, 179–180
  natural gas, 102, 198–199
  oxychlorination, 198
  R&D, 186
  synthesis gas, 87–88, 90
Ceramics, 61, 184
Chemical processes
  coal-oil coprocessing, 98–99, 127
  heavy oil, 69
  oil and gas recovery, 30–31, 35–36
  oil shales, 85–86
  photochemical oxidation, 107–108, 110–111
  tar sands, 3, 44
  see also specific processes
Chem Systems, Inc., 88
Chevron, 82, 98, 184
Chlorofluorocarbons, 15, 111
Clean Air Act, 105–106
Climate and weather
  air quality models, 107, 110–111
  photochemical oxidation, 107–108, 110–111
  see also Greenhouse effect
Coal, 119, 121–122, 125–127
  commercial applications, 61–62, 63, 73, 92, 96–97, 126, 200
  consumption, 15
  coprocessing with oil, 7, 97–100, 127, 188–190
  cost factors, 7, 44, 60–62, 64–65, 91–92, 93, 94, 95–96, 99–100, 121–122, 127
  DOE programs, 7, 8, 61, 63–65, 93, 95, 96–98, 100–102, 121, 126–127, 129
  environmental factors, 61–62, 63, 65, 92, 94, 99, 101
  gasification, 3, 13, 44, 58–62, 63, 65, 89, 120, 122, 200
  liquefaction, 3, 4, 7, 11, 12, 44, 45, 61, 64–65, 92–97, 99–100, 115, 120, 121–122, 127, 188
  methanol production, 44–45, 91
  prices, 42, 92, 93
  pyrolysis, 64, 100–102, 129, 191–196
  R&D, 7, 11
  resources, 16, 17, 92, 126, 145

INDEX

underground gasification, 3, 44, 63, 65, 91–92
Coke and coking
  coal pyrolysis, 64, 100–102, 129, 191–196
  heavy oil, 66–67, 68, 69, 178–179
  tar, 74
Colorado, 17, 76
Commercial applications, 5, 116–117, 123, 125, 133
  asphalt additives, 13
  auto engines, 111; catalytic converters, 107–108
  coal gasification, 61–62, 63, 73, 200
  coal, general, 126
  coal liquefaction, 92, 96–97
  coprocessing, 188–190
  heavy oil, 66–67, 178–182
  methanol-to-gasoline, 88–89
  natural gas-to-liquids F-T facility, 89–90
  oil shale, 11, 13, 76–79, 86–87, 126, 143, 183–184
  RD&D, 13, 97, 121, 185–187
  synthesis gas, 89–90
  tar, 71–74
Compressed natural gas (CNG), 1, 3, 119
  cost factors, 40, 41–42, 45, 53, 54–56, 108–109, 150
  environmental factors, 107–109, 110, 111–113
Consumers and consumption, 2, 119
  coal, 15
  natural gas, 15, 145
  oil, 13, 14, 139
Controlled Retracting Injection Point technology, 63
Cool Water demonstration project, 61
Coprocessing, 7, 62, 64–65, 97–100, 127, 188–190
Cost factors, 1, 3–4, 5, 6, 116, 117, 120, 149
  alternative fuels production, general, 40–56, 146–177
  auto operating, 49, 50–51, 53–55
  coal gasification, 60–62, 64, 91–92, 122

coal liquefaction, 7, 44, 61, 64–65, 92, 93, 94, 95–96, 99–100, 121–122, 127
coal-oil coprocessing, 99–100
coal pyrolysis, 101
compressed natural gas, 40, 41–42, 45, 53, 54–56, 108–109, 150
crude oil, 121
discount rate, 150–152, 156–172
environmental factors, 106, 118
ethanol, 45, 53
gasoline, 149
heavy oil, 68
methanol, 40, 42, 43, 44–45, 48–52, 53–56, 90–91, 120, 149, 154, 175–177
methodology, 20, 40–53, 146–177
natural gas, 48, 58, 168–177
natural gas liquids, 27, 103–104
oil, enhanced recovery, 34, 118
oil shale, 44, 80, 83–84, 86–87, 121, 126
R&D and, 56
synthesis gas, 90–91
tar sands, 44, 71, 75, 76
see also Capital investments; Price factors
Cracking
  fluid, 66–67, 68, 179–180
  hydro, 67, 100
  thermal, 67, 100
  see also Catalysis and catalytic cracking
Crude oil, 10, 139–141
  cost, 121
  prices, 15, 18, 46–48, 56, 58, 147–149
  production, 6, 13–14, 15, 22, 28–31, 32–33, 37, 126
  see also Tar and tar sands

D

Department of Energy (DOE), 4, 5–9, 17, 20, 116, 117, 123–129, 134
  biomass resources, 17, 65, 66, 126, 128–129

coal, 126-127
coal gasification, 61, 63-65
coal liquefaction, 7, 64, 65, 93, 95, 96-97, 121
coal-oil coprocessing, 7, 98, 100, 127
coal pyrolysis, 8, 64, 101-102, 129
conversion technologies, 20, 64-65
environmental R&D, 7, 110-111, 122-123
heavy oil, 8, 65, 68-69, 128
methane, 8-9, 129
methanol, 8, 104, 129
oil and natural gas, 6, 8-9, 38, 65, 104, 125-126, 128, 129
oil shale, 9, 82, 84-86, 126-127, 129, 183
spending, 11-12, 38, 123-124
synthesis gas, 8, 61, 63, 64-66, 88, 90, 92, 120
tar, 7, 8, 75-76, 128
Department of the Interior, 25-27
Diesel fuel, 3, 13
environmental impacts, 108-109
methanol conversion, 88-89
Diseases and disorders, 94, 109-110
Distillate fuel, 94
middle distillate synthesis, 42, 44, 48, 90
Distribution issues, alternative fuels, 53, 55-56, 91, 104, 149-150
Dow Chemical, 13, 90

E

Ebullating bed reactor, 67, 181
Economic factors, 2-3, 120, 133
alternative fuels, 40-57
biomass, 145
coal-oil coprocessing, 99, 100, 127
coal processes, general, 65-66
coal pyrolysis, 101
funding, government, 6-9, 93-94, 96, 125-127; see also Department of Energy
natural gas, 23, 103-104
oil and gas, 23, 32-33
oil shale, 65-66, 77, 84-85

planning, 18, 151
royalties, 42
synthesis gas, 58-60, 64-65, 90-92
tar sands, 75
see also Capital investments; Commercial applications; Consumers and consumption; Cost factors; Price factors; Production and productivity; Resource base
Efficiency
methanol- *vs.* gasoline-fueled autos, 49, 50-51, 53-56, 149
natural gas technology, 144
Equipment, oil rigs, 34
see also Capital investments
Electric power, 2, 6, 119
coal gasification, 61
conversion technologies and, 65, 89
methanol, 89
tar, 74, 75
for transportation, 13
Electric Power Research Institute, 95
Electrolysis, 4
End-use factors, 7, 53-55
Energy Information Administration (EIA), 14, 18, 38, 148-149
Enhanced oil recovery, 29-30, 33, 33-34, 35-36, 37, 118
Environmental Protection Agency, 105, 107, 110, 127
Environment and pollution, 1, 5, 6, 7, 14, 19, 105-114, 116, 122-123, 127
ash, 62, 95, 101
biomass energy, 113-114, 128
carbon monoxide, 106, 119
coal gasification, 61-62, 63, 65
coal liquefaction, 92, 94
coal-oil coprocessing, 99
coal pyrolysis, 101
compressed natural gas, 107-109, 110, 111-113
cost and, 106, 118
DOE programs, 7, 110-111, 122-123
formaldehyde, 107, 108, 110
heavy oil, 67, 68, 69, 128

methanol, 107, 108, 109–110
oil and natural gas production, 31–32
oil shale, 78–79, 82–83
ozone, 105, 106, 107, 110, 111
photochemical oxidation, 107–108, 110–111
regulations and standards, 2, 19, 105–108, 122
tar sands, 73, 74–75, 76, 128
urban areas, 5, 13, 14, 104, 105, 106, 110, 122
volatile organic compounds, 105
see also Carbon dioxide; Climate and weather; Greenhouse effect; Nitrogen and nitrogen oxides; Sulfur
Ethanol, 45, 53
Ethylene, 198–199
Exploration, 2
defined, 185
oil and natural gas, 23–24, 31, 32, 37
oil shale, 84
Exports and imports, *see* Imports and exports
Exxon, 11, 90, 199
Exxon Valdez, 31

F

Federal government
alternative-fuel vehicle demonstrations, 105
Australia, 51–52, 93–94
funding, 6–9, 93–94, 96, 125–127
Germany, Federal Republic of, 94
Japan, 93–94
laws, 105–106
oil shale ownership, 84
regulations and standards, 2, 19, 105–108, 122
spending, 11–12, 38, 93–94, 96, 123–124
taxes, 116, 118
see also Department of Energy; Department of the Interior; Environmental Protection Agency

Fischer assay, 77
Fischer-Tropsch process, 57, 89–90, 91, 92, 120, 123, 129
Fluid catalytic cracking, 66–67, 68, 179–180
Formaldehyde, 107, 108, 110
Fossil fuels
cost factors, 3–4
R&D, 11
see also specific fuels
France, 97
Fuel properties
coal, liquefied, 94, 96
coal-oil coprocessing, 99
coal pyrolysis, 101
heavy oil, 67–68
oil shale,
Funding, government, 6–9, 93–94, 96, 125–127
see also Department of Energy

G

Gas and gasification,
biomass, 65, 66
coal gasification, 3, 13, 44, 58–62, 63, 65, 89, 120, 122, 200
underground coal gasification, 3, 44, 63, 65, 91–92
see also Carbon dioxide; Compressed natural gas; Hydrogen; Natural gas; Nitrogen and nitrogen oxides; Synthesis gas
Gasoline, 3, 13, 46, 149
coal to, 94, 95
methanol to, process, 43, 48, 57, 88–89, 91, 103, 123, 149
multifuel engines, 107
natural gas to, 3, 89
prices, 46, 149
Geokinetics, 184
Geological Survey, 26, 27
Geology and geophysics
advanced monitoring, 37
modeling, 23, 25, 35
oil and gas, 21–24, 31
oil shale, 76–78

tar sands, 69–70
see also Exploration; Resource base
Germany, Federal Republic of, 88, 94, 97, 98
Government, *see* Federal government; State government
Greenhouse effect, 1–2, 4, 15, 19, 111–114, 116, 122
Gulf-Badger, 90
Gulf Research and Development, 63, 184

H

Health issues, *see* Diseases and disorders; Environment and pollution
Heavy metals, 67, 68, 98, 128, 190
Heavy oil
commercial applications, 66–67, 178–182
coprocessing with coal, 7, 97–100, 127, 188–190
DOE programs, 8, 65, 68–69, 128
environmental factors, 67, 68, 69, 128
resources, 16, 117–119, 141–142
High-temperature processes, *see* Coke and coking; Pyrolysis; Retorting; Thermal processes
Historical perspectives
coal liquefaction, 92
coal pyrolysis, 100
crude oil *vs.* refined gasoline, prices, 46
hydrogen production, 60–61
oil and gas resources, 24–25
Hydrocarbon Research, Inc., 89
Hydrocracking, 67, 100
Hydrogen, 57–63, 66, 67, 69, 94, 113, 119, 120, 123, 126, 128, 180–182

I

ICF Resources, Inc., 19, 25–27
Idemitsu Kosan, 198
Imports and exports, 2, 10, 11, 13–14, 103–104, 115, 118

Integrated gasification combined cycle, 89
Intellectual property, 118–119
International Energy Agency, 97
International perspective
coal liquefaction, 97
Greenhouse effect, global cooperation, 15
hydrocarbon resource base, 138–145
investment, 2, 118
natural gas, 58, 91, 103–104
see also Imports and exports; specific countries
Iron and iron oxide, 62–63, 82
Italy, 97

J

Japan, 63, 68–69, 93–94, 97, 198, 199

K

Kentucky, 16, 93, 143, 190
Kerogen, 77–78, 85
Kerr-McGee, 13, 98

L

Law
Clean Air Act, 105–106
intellectual property, 118–119
regulations and standards, 2, 19, 105–108, 122
taxes, 116, 118
Lawrence Livermore National Laboratory, 82, 184
Leasing, 32
Licenses, 118–119
Liquefied natural gas, 27–28, 89, 139, 140

M

Malaysia, 90
Methane, *see* Natural gas
Methanol, 1, 3–4, 8, 89, 120, 129, 150
coal to, 63

cost factors, 40, 42, 43, 44–45, 48–52, 53–56, 90–91, 120, 149, 154, 175–177
DOE programs, 8, 104, 129
environmental factors, 107, 108, 109–110
Fischer-Tropsch process, 89–90, 120, 123, 129
to gasoline process, 43, 48, 57, 88–89, 91, 103, 120, 149
natural gas to, 45, 48–49, 58, 102, 197
to olefins process, 88–89
from synthesis gas, 87–90, 92, 129
Methodology, 20, 40–53, 146–177
Middle distillate synthesis, 42, 44, 48, 90
Middle East, 91, 92, 103, 133, 139, 145
Organization of Petroleum Exporting Countries, 13, 18, 22
Minerals Management Survey, 26, 27
Mining
coal, 92
oil shale, 76, 78–79, 86
tar sands, 71, 74–75, 76, 128
Miscible extraction methods, 30, 35
Mitsubishi Gas Chemical, 87–88
Mobil Oil, 88–89, 90, 103
Models and modeling
air quality, 107, 110–111
geological reservoirs, 23, 25, 35
oil shale processes, 83
Morgantown Energy Technology Center, 65
Multifuel systems, 53–54, 107, 112, 149

N

National Institutes of Health, 110, 127
National security, 13, 19
Naptha, 73, 89, 94
Natural gas, 8–9, 13, 15, 21–39, 62, 120, 125–126, 129
consumption, 15, 145
conversion processes, 58, 65, 102–104, 197–199

cost factors, 27, 48, 58, 103–104, 168–177
DOE programs, 8–9, 129
environmental factors, 31–32
gasoline produced from, 3, 89
liquids, 27–28, 89, 139, 140
methanol from, 45, 48–49, 58, 102, 197
prices, 3, 4, 14, 48, 66, 103, 115
production, 6, 14, 28–31, 32–33, 115
resources, 16, 19, 21–25, 27–28, 115, 119–120, 138–139, 140, 144–145
synthesis gas and, 91, 200
see also Compressed natural gas
New Paraho Shale Oil Company, 13
New Zealand, 88
Nitrogen and nitrogen oxides
auto emissions, general, 106, 107, 110–112
coal-oil coprocessing, 190
compressed natural gas, 107–108
heavy oil, 67, 68, 69, 128
natural gas, 197
oil shale, 82
synthesis gas, 60
tar, 73, 74, 75, 76, 128
North Dakota, 61

O

Occidental Petroleum, 13, 76, 79, 194
Occidental Research Corporation, 101, 183
Occupational health and safety, 94
Office of Conservation and Renewable Energy, 11
Office of Fossil Energy, 11, 12
Office of Program Analysis, 96
Ohio, 98
Oil, 21–39
coal-oil coprocessing, 7, 97–100, 127, 188–190
consumption, 13, 14, 139
enhanced oil recovery, 29–30, 33, 33–34, 35–36, 37, 118
environmental factors, 31–32

prices, 3–4, 5–6, 13–15, 18, 22, 24, 39, 46–48, 56, 58, 60–61, 65–66, 115, 116, 117–118, 120, 123, 148–149
production, 6, 14, 28–31, 32–33
R&D, 11
residuum, 8, 128, 179–181
resources, 14, 15, 16, 19, 24–27, 33–34, 115, 117–119, 138, 139–141
see also Crude oil; Heavy oil; Tar and tar sands
Oil shale, 76–87, 121–122, 120, 122, 129
  commercial applications, 11, 13, 76–79, 86–87, 126, 143, 183–184
  costs, 44, 80, 83–84, 86–87, 121, 126
  DOE programs, 9, 82, 84–86, 126–127, 129, 183
  environmental factors, 78–79, 82–83
  pyrolysis, 44, 78
  kerogen, 77–78, 85
  R&D, 7, 9, 11, 82, 84–87, 115, 119, 126–127, 183–184
  resources, 16, 17, 76, 126, 143
  retorting, 11, 13, 78, 79–80, 80–84, 85, 87, 183–184
Olefins, methanol to, process, 88–89
Organization of Petroleum Exporting Countries, 13, 18, 22
Oxychlorination, 198
Oxygen and oxidation processes, 60, 61–62, 90, 102, 103, 197, 198–199
  photochemical oxidation, 107–108, 110–111
  REDOX process, 102, 103, 199
Ozone, 105, 106, 107, 110, 111

P

Petroleum, *see* Oil
Phillips Petroleum, 197, 199
Photochemical oxidation, 107–108, 110–111
Pipelines
  methanol, 91
  oil shale, 78
Planning, 1, 4, 5–9, 17–19, 97, 115–129, 151
  see also Department of Energy
Pollution, *see* Environment and pollution
Potential Gas Committee, 27
Price factors, 2, 3, 4, 5, 11, 120
  coal, 42, 92, 93
  gasoline *vs.* crude oil, 46, 149
  methanol, 103
  methodology, 41, 42–43, 46–53, 146, 147–148
  natural gas, 3, 4, 14, 48, 66, 103, 115
  oil and natural gas, 3–4, 5–6, 13–15, 18, 22, 24, 39, 46–48, 56, 58, 60–61, 65–66, 115, 116, 117–118, 120, 123, 148–149
  tar, 75, 128
  volatility, 2, 18, 22, 60–61, 123
Private sector 5, 11, 13
  oil industry, 23, 38
  R&D, 5, 11, 13
  tar, 128
  see also Commercial applications; *specific companies*
Production and productivity, 2, 5
  alternative fuels, costs, 40–56
  coal and oil shale, 7
  hydrogen, 57–63
  oil and gas, 6, 13–14, 15, 22, 28–31, 32–33, 37, 115, 126
  synthesis gas, 57–63
  see also Efficiency; Resource base
Projections
  air quality, 14
  methodology, 25–27, 41, 42–43, 46–53, 147–148
  oil shale, 87
  planning, 1, 4, 5–9, 17–19, 35–39, 97, 115–129, 151
  prices, 18, 117
  see also Greenhouse effect; Resource base

Pyrolysis, 3, 4, 8, 57, 199
  coal, 64, 100–102, 129, 191–196
  Fischer assay, 77
  natural gas, 199
  oil shale, 44, 78
  tar sands, 3, 44
  see also Coke and coking

R

REDOX process, 102, 103, 199
Regulations and standards, 2, 19, 105–108, 122
Research and development, 10–13, 17, 115–129, 133–134
  academic research, 69, 98–99, 197–198, 199
  auto engines, 53–54, 106–114
  biomass, 4, 8, 65, 66, 126, 128–129
  capital costs and planning, 151–154
  coal, 119, 125–127
  coal gasification, 57–66, 200
  coal liquefaction, 4, 64, 65, 92–94, 96–97, 115, 121–122
  coal-oil coprocessing, 7, 97–100, 127, 188–190
  coal pyrolysis, 8, 64, 100–102, 129, 191–196
  commercial applications and, 116–177
  conversion technologies, general, 20, 57–104, 197–199
  cost of production and, 56
  environmental issues, 105–114, 122–123
  fossil fuels, general, 11
  heavy oils, 68–69, 178–182
  hydrogen production, 57–63
  methodology, 20, 40–53, 146–177
  natural gas, 58, 104, 125–127
  oil and natural gas, 11, 35–39, 125–126
  oil shale, 7, 9, 11, 82, 84–87, 115, 119, 126–127, 183–184
  planning, 1, 4, 5–9, 17–19, 35–39, 97, 115–129, 151
  synthesis gas, 57–66, 87–92

RD&D, 13, 97, 121, 185–187; *see also* Commercial applications
  tar sands, 4, 75–76
  trends, 2, 3
  see also Department of Energy; Exploration
Research Association for Residual Oil Processing, 69
Research Association of Petroleum Alternative Development, 69
Resource base, 16, 138–145
  biomass, 17, 128, 138, 145
  coal, 16, 17, 92, 126, 145
  enhanced oil recovery, 29–30, 33, 33–34, 35, 37, 118
  estimate methods, 25–27
  heavy oils, 16, 117–119, 141–142
  hydrocarbons, 138–145
  natural gas, 16, 19, 21–25, 27–28, 115, 119–120, 138–139, 140, 144–145
  natural gas liquids, 27–28, 139, 140
  oil, 14, 15, 16, 19, 24–27, 33–34, 115, 117–119, 138, 139–141
  oil shale, 16, 17, 76, 126, 143
  tar, 16, 70, 117–119, 141–143
  see also Production and productivity
Retorting
  kerogen, 78, 85
  oil shale, 11, 13, 78, 79–80, 80–84, 85, 87, 183–184
  tar sands, 71, 73, 75
Royalties, 42

S

Saudi Arabia, 13
SFA Pacific, Inc., 20
Shell middle distillate synthesis, 42, 44, 48, 90, 103
Shell Oil, 13, 44, 90
Sohio, 199
South Africa, 89–90
Soviet Union, 63, 139, 141, 145
Standards and regulations, 2, 19, 105–108, 122
State government, 105, 106, 107
  taxes, 116

STATOIL, 90
Sulfur
  coal-oil coprocessing, 190
  coal pyrolysis, 101
  heavy oil, 67, 68, 69, 128, 180–181
  oil shale, 78, 82
  synthesis gas, 65
  tar, 70, 73, 74, 75, 128
Surface Coal Gasification Program, 102
Synthesis gas, 8, 45, 57–66, 87–92, 103, 119–120, 200
  carbon dioxide and, 27, 103–104
  catalysis, 27, 103–104
  commercial applications, 89–90
  cost factors, 90–91
  DOE programs, 8, 61, 63, 64–66, 88, 90, 92, 120
  economic factors, 58–60, 64–65, 90–92
  Fischer-Tropsch process, 57, 89–90, 91, 92, 120, 123, 129
  methanol from, 87–90, 92, 129
  natural gas and, 91, 200
  nitrogen oxides and, 60
  sulfur and, 65

T

Taiwan, 199
Tar and tar sands, 69–76, 117–119, 120
  cost factors, 44, 71, 75, 76
  DOE programs, 7, 8, 75–76, 128
  environmental factors, 73, 74–75, 76, 128
  prices, 72, 128
  pyrolysis, 3, 44
  resources, 16, 70, 117–119, 141–143
  solvent extraction, 3, 42, 44, 71–72, 75, 76
  see also Bitumen
Taxes, 116, 118
Technological innovation
  transfer of, 6, 38, 118–119, 128
  see also Alternative fuel sources; Commercial applications; Research and development
Texaco, 13, 61

Texas, 16, 88, 93, 143
Thermal processes
  coal gasification, 61–63, 65, 200
  coal liquefaction, 92–94, 95
  coal-oil coprocessing, 98
  cracking, 67, 73, 90
  enhanced oil recovery, 29–30, 36–37
  high-temperature gas-cooled reactor, 200
  natural gas conversion, 58
  synthesis gas, methanol production, 87–90, 129
  tar conversion, 73
  see also Coke and coking; Greenhouse effect; Pyrolysis; Retorting
Time factors, 5, 116
  committee meeting schedule, 135–137
  oil and gas production, 32–33
  oil shale technology, 86–87
Toxicity, 109–110
Trade, *see* Imports and exports
Transportation systems, specific
  air, 13
  coal costs, 92
  distribution of alternative fuels, 53, 55–56, 91, 104, 149–150
  pipelines, 78, 91
  problems, 13–15
  see also Automobiles

U

Uhde Gmbh, 88
Underground coal gasification, 3, 44, 63, 65, 91–92
Union Carbide, 90, 198
Union Oil, 11
Union Rheinische Braunkohlen Kraftstoff AG, 88
United Kingdom, 90, 92, 94
Universities, *see* Academic research
University of California, 197–198, 199
University of Houston, 199
University of Manitoba, 197
University of Michigan, 199
University of Pittsburg, 198

University of Taiwan, 199
University of Tokyo, 198
Urban areas, 5, 13, 14, 104, 105, 106, 110, 122
Utah, 16, 17, 74, 76, 77, 82

V

Vacuum residue, 180–181
Veba Oil, 98
Venezuela, 70, 139, 143

Viscosity, 78, 80, 141
Volatile organic compounds, 105

W

Washington State, 92, 92
Water pollution, *see* Environment and pollution
Weather, *see* Climate and weather
Western Research Institute, 76
Wood and wood products, 44–45, 111
Wyoming, 17, 76, 77